Theoretical Biology

Series Editor

Yoh Iwasa, Department of Bioscience, School of Science and Technology, Kwansei Gakuin University, Hyogo, Japan

The “Theoretical Biology” series publishes volumes on all aspects of life sciences research for which a mathematical or computational approach can offer the appropriate methods to deepen our knowledge and insight.

Topics covered include: cell and molecular biology, genetics, developmental biology, evolutionary biology, behavior sciences, species diversity, population ecology, chronobiology, bioinformatics, immunology, neuroscience, agricultural science, and medicine.

The main focus of the series is on the biological phenomena whereas mathematics or informatics contribute the adequate tools. Target audience is researchers and graduate students in biology and other related areas who are interested in using mathematical techniques or computer simulations to understand biological processes and mathematicians who want to learn what are the questions biologists like to know using diverse mathematical tools.

More information about this series at http://www.springer.com/series/15703

Akihiko Mougi
Editor

Diversity of Functional Traits and Interactions

Perspectives on Community Dynamics

Springer

Editor
Akihiko Mougi
Faculty of Life and Environmental Sciences
Shimane University
Matsue, Shimane, Japan

ISSN 2522-0438 ISSN 2522-0446 (electronic)
Theoretical Biology
ISBN 978-981-15-7952-3 ISBN 978-981-15-7953-0 (eBook)
https://doi.org/10.1007/978-981-15-7953-0

This Springer imprint is published by the registered company Springer Nature Singapore Pte Ltd.
The registered company address is: 152 Beach Road, #21-01/04 Gateway East, Singapore 189721, Singapore

Preface

The goal of this book is to present a new theoretical perspective of ecological community dynamics. A classical theory defined complexity in ecological communities by the numbers of species and interaction links, and predicted that complex communities with diverse species and their interactions between them are inherently unstable or not likely to persist, contrary to natural communities. Since then, tremendous number of theoretical studies have proposed key mechanisms for stabilizing or regulating community dynamics in nature. Those earlier studies have mainly focused on a specific characteristic of communities, architecture of species interaction network such as interaction strengths and network topologies. However, beyond such classical simple node-link network view, recent studies have developed theories including more realistic community structures with the elements of biodiversity itself, diversity of intraspecific and interspecific traits.

Real ecological communities are not mere interaction network comprised of nodes and links, which respectively represent species and interactions. Nodes in ecological communities differ each other reflecting traits and characteristics of each species and habitat and that links in ecological communities are not characterized by the simple presence or absence and interaction strengths, but rather by diverse interaction types and functional responses. This book proposes that diversity of traits and interactions is a key to understand the dynamics of ecological communities. Furthermore, it introduces innovative methods to understand such very complex ecological communities.

Matsue, Japan

Akihiko Mougi

Contents

Part I
Intra-Specific Trait Diversity

Chapter 1
Incorporating Ontogenetic Niche Shifts into Hybrid Community Dynamics

Takefumi Nakazawa

Abstract Many organisms on the Earth are multicellular, and thus typically change their niches (e.g., behaviors and diets) during growth (i.e., ontogenetic niche shifts). Previous theoretical studies have shown that ontogenetic diet shifts undermine the stability of stage-structured food webs, wherein all interactions are assumed to be antagonistic (i.e., predator–prey). Meanwhile, real ecological communities are formed by different types of interactions, including not only antagonistic predation but also mutualism and competition (i.e., hybrid community). At present, however, it remains unclear how ontogenetic niche shifts mediate the stability of "stage-structured hybrid community," wherein not only diet but also interaction type may change because of ontogenetic niche shift. To fill the knowledge gap, I overview the integration of stage-structured food-web theory into theory of hybrid community. Specifically, I will model the novel situation that plants interact with antagonistic herbivores and mutualistic pollinators/seed dispersers at the vegetative and reproductive stages, respectively, while introducing my previous study for the situation that insects act as antagonistic herbivores and mutualistic pollinators at the larval and adult stages, respectively. Notably, the results predict that such ontogenetic changes in interaction types between antagonism and mutualism can increase the stability of stage-structured communities, beyond the previous concern regarding the instability of stage-structured food-web models.

Keywords Antagonism–mutualism · Hybrid community dynamics · Life history · Plant–insect interaction · Herbivory–pollination · Ontogenetic niche shift · Stage structure

---

T. Nakazawa (✉)
Department of Life Sciences, National Cheng Kung University, Tainan, Taiwan

A. Mougi (ed.), *Diversity of Functional Traits and Interactions*, Theoretical Biology,
https://doi.org/10.1007/978-981-15-7953-0_1

1.1 Introduction

Many organisms on Earth are multicellular and thus exhibit an ontogenetic increase in body size. Further, during individual growth, these organisms often change their diets or habitats qualitatively or quantitatively, which are termed as ontogenetic niche shifts (Wilbur 1980; Werner and Gilliam 1984). Traditionally, research on community ecology and biodiversity has ignored the role of ontogenetic niche shifts in shaping community structure, under the simplified assumption that species comprise identical individuals with invariant traits (Begon et al. 2006; Verhoef and Morin 2010). It is only recently that researchers have recognized the importance of addressing how ontogenetic niche shifts can mediate community dynamics through changes in interspecific interactions, thereby influencing biodiversity and ecosystem functioning (reviewed by Miller and Rudolf 2011; de Roos and Persson 2013; Nakazawa 2015).

Notably, recent theoretical work predicts that ontogenetic niche shifts can critically affect community dynamics. In particular, they can decrease the community resilience. To take a simple example, consider a one consumer-two resource system with a juvenile–adult stage structure in the consumer, in which the consumer feeds on different resources at the juvenile and adult stages. Theoretical models predict that the dynamics of such a stage-structured community will converge to either a juvenile-dominated or an adult-dominated state, depending on the initial conditions (Schreiber and Rudolf 2008; Guill 2009; Nakazawa 2011a, 2011b; also see below for the mechanism). Considering the ubiquity of ontogenetic niche shifts among organisms, the prediction implies that alternative community states may commonly occur over a range of environmental conditions. Further, it is also suggested that community structure may drastically change following an environmental perturbation (regime shifts), and once the community is trapped in another stable state, it is difficult to recover the original state via manipulation of environmental conditions (hysteresis). This idea is critical for resilience-based ecosystem management (Nakazawa 2014) and biodiversity research (Mougi 2017).

However, it should also be pointed out that previous theoretical studies of stage-structured community dynamics have considered only antagonistic systems (i.e., ontogenetic diet shifts of predators in food webs), whereas real communities include different interaction types, including not only predation but also mutualism and competition (i.e., hybrid community). Therefore, our present understanding of the community-level consequences of ontogenetic niche shifts remains incomplete.

1.2 Toward Stage-Structured Hybrid Community

Another branch of research in community ecology, the hybrid community theory, has recently highlighted the role of interaction-type diversity in community dynamics (Mougi and Kondoh 2012; Sauve et al. 2014; Kondoh and Mougi 2015; Lurgi et al. 2016). The hybrid community theory predicts that even if a purely antagonistic (food-web) or mutualistic system is not persistent, their combination can be stable and enable many species to coexist therein (Mougi and Kondoh 2012). This may resolve the long-standing debate over the diversity–stability relationship (May 1972; Mcnaughton 1978; Pimm 1984; Ives and Carpenter 2007; Allesina and Tang 2012, 2015), which emerged because antagonistic (food-web) and mutualistic systems have traditionally been studied independently (see independent publications of McCann 2011; Bascompte and Jordano 2014). The idea of hybrid community is appealing because interactions among the most diverse organisms on Earth (i.e., insects and plants) are actually a hybrid of antagonism and mutualism (i.e., herbivory and pollination, respectively). However, even the hybrid community theory has also overlooked the common fact that many organisms are stage structured, and ontogenetic niche shifts can critically affect stage-structured community dynamics (see above).

Taken together, it is apparent that incorporating ontogenetic niche shifts (stage structure) into the hybrid community theory is needed to establish the two key theories (i.e., stage-structured food-web dynamics and nonstructured hybrid community dynamics), which would help improve our understanding of community dynamics and biodiversity mechanisms. This novel community view may be called a "stage-structured hybrid community," in which organisms change not only the interacting species (e.g., resources and predators) but also the interaction type, because of ontogenetic niche shifts. Note that stage-structured hybrid communities are commonly observed in interactions between insects and plants, the most diverse organisms. A representative example is the metamorphosis of pollinating insects like Lepidopteran (butterflies and moths), in which larvae are antagonistic herbivores whereas adults are mutualistic pollinators (Altermatt and Pearse 2011). Similar examples are observed in other holometabolous insects, major pollinators, including Hymenoptera (e.g., bees), Diptera (e.g., flies), and Coleoptera (e.g., beetles) (Wäckers et al. 2007), in which the larvae do not contribute to pollination but just exploit the plant resources. In addition, stage structures in plants can also result in stage-structured hybrid communities, because plants are attacked by herbivores at the vegetative stage, whereas they often require pollinators at the reproductive stage. Although extensive research has been carried out to understand the nature of herbivory and pollination, very few of these studies have considered the role of both interaction types across the plant life history (Bronstein et al. 2007; Boege et al. 2011; Ikemoto et al. 2017).

In this chapter, I briefly overview the major dynamical features of simple community modules for stage-structured hybrid communities, while addressing the novel community view. I specifically highlight two distinct types of community modules as introduced above, depending on the species that exhibit the stage structure. One is a consumer-mediated ontogenetic antagonism–mutualism coupling (Fig. 1.1a), in which juveniles and adults of a consumer species utilize different resources as antagonists and mutualists, respectively, such as holometabolous pollinating insects (Ke and Nakazawa 2018). The other is a resource-mediated ontogenetic antagonism–mutualism coupling (Fig. 1.1b), in which juveniles and adults of a

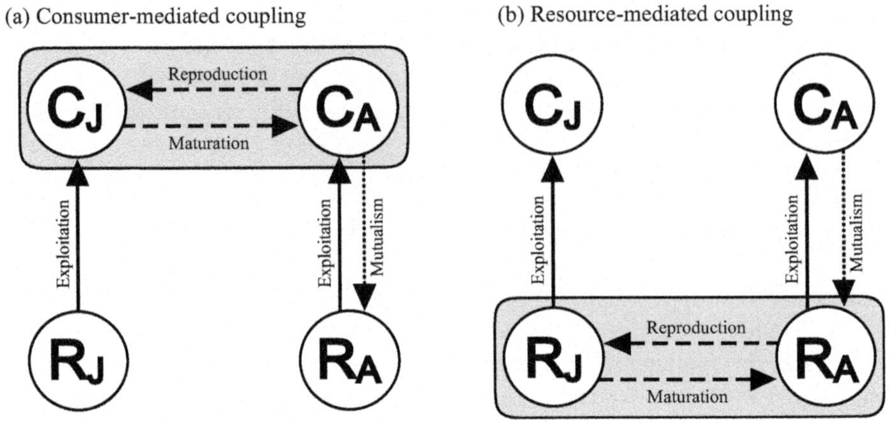

Fig. 1.1 Conceptual module diagrams of (**a**) consumer-mediated and (**b**) resource-mediated ontogenetic antagonism–mutualism coupling (adapted from Ke and Nakazawa 2018). (**a**) Antagonistic juveniles C_J and mutualistic adults C_A utilize different resources R_J and R_A, respectively. (**b**) The resource interacts with antagonistic C_J and mutualistic C_A consumers at the juvenile R_J and adult R_A stages, respectively. Gray boxes represent the species identity in which maturation and reproduction couple the juvenile and adult stages. Interspecific interactions denoted by solid and dotted lines indicate resource exploitation (e.g., herbivory and pollen/nectar feeding) and mutualism (e.g., pollination), respectively

resource species interact with an antagonist and a mutualist, respectively, such as flowering plants. Finally, I discuss future perspectives to extend the modules with more realistic descriptions.

1.3 Consumer-Mediated Ontogenetic Antagonism–Mutualism Coupling

Consider a consumer species that exploits a resource at the juvenile stage, while acting as a mutualist of another resource at the adult stage (Fig. 1.1a). The community module for this situation can be formulated as follows (Ke and Nakazawa 2018):

$$\frac{dR_J}{dt} = r_J\left(1 - \frac{R_J}{K_J}\right)R_J - F_J(R_J)C_J \tag{1.1a}$$

$$\frac{dR_A}{dt} = r_A\left(1 - \frac{R_A}{K_A}\right)R_A - F_A(R_A)C_A + B(C_A)R_A \tag{1.1b}$$

$$\frac{dC_J}{dt} = b_AF_A(R_A)C_A - b_JF_J(R_J)C_J - d_JC_J \tag{1.1c}$$

$$\frac{dC_A}{dt} = b_J F_J(R_J) C_J - d_A C_A, \tag{1.1d}$$

where R_i ($i = J$ or A) is the stage-specific resource abundance and C_i is the juvenile or adult abundance of the consumer. The resources exhibit a logistic growth, with intrinsic growth rate r_i and carrying capacity K_i. The juveniles and adults of the consumer utilize different resources to mature to adulthood and reproduce offspring, respectively. The function $F_i(R_i)$ is the functional response of stage-specific resource exploitation, and the parameter b_i is its conversion efficiency. Note that the adults are mutualists, but they also exploit the resource, such as pollinating insects feeding on nectar or pollen of flowering plants; this is called the consumer–resource approach to mutualism (Holland and DeAngelis 2010). The adults facilitate the population growth of their resource R_A, according to the function $B(C_A)$. The parameter d_i is the stage-specific death rate of the consumer.

Here, all the interaction terms are assumed to be linear (Holling's type-I) for analytical tractability; that is, $F_i(R_i) = a_i R_i$ and $B(C_A) = \alpha C_A$, where parameter a_i ($i = J$ or A) is the efficiency of stage-specific resource exploitation (i.e., herbivory or nectar/pollen feeding) and parameter α is the efficiency of adult mutualism (i.e., reproductive success of flowering plants). Importantly, for $a_A > \alpha$, the model (1) is identical to the corresponding food-web module, in which the consumer is antagonistic at both juvenile and adult stages (Schreiber and Rudolf 2008). Therefore, I compare the dynamics in situations when $a_A > \alpha$ (i.e., the adults are antagonists) and when $\alpha > a_A$ (i.e., the adults are mutualists) to illustrate the importance of ontogenetic antagonism–mutualism coupling.

Consider the coexistence equilibrium where the growth rates of all variables are zero (i.e., $dR_i/dt = 0$ and $dC_i/dt = 0$). From $dRi/dt = 0$, the equilibrium resource abundances are solved as $R_J^* = K_J(1 - a_J C_J^*/r_J)$ and $R_A^* = K_A[1 - (a_A\text{-}\alpha)C_A^*/r_A]$, respectively. Using these relationships, the four equilibrium equations of the model (1) are reduced to the following two relations between the juvenile and adult equilibrium abundances (Ke and Nakazawa 2018):

$$C_A^* = \frac{b_J a_J K_J}{d_A}\left(1 - \frac{a_J}{r_J} C_J^*\right) C_J^* \tag{1.2a}$$

$$C_J^* = \left[\frac{b_A a_A K_A}{d_J}\left(1 - \frac{a_A - \alpha}{r_A} C_A^*\right) - \frac{d_A}{d_J}\right] C_A^*. \tag{1.2b}$$

These are quadratic functions of each other, and their intersections determine the equilibrium points. Obviously, Eq. (1.2a) is always an upward convex function, with its vertex in the first quadrant on the C_J^*–C_A^* plane (solid curves in Fig. 1.2). In contrast, the functional form of Eq. (1.2b) changes, depending on whether $a_A > \alpha$ (Figs. 1.2a–c) or $\alpha > a_A$ (Figs. 1.2d–f). This implies that whether the adults are antagonists or mutualists has dynamical consequences.

If the adults are antagonists ($a_A > \alpha$), Eq. (1.2a) is a rightward convex (dashed curves in Figs. 1.2a–c). In this scenario, when either of the two resource-carrying

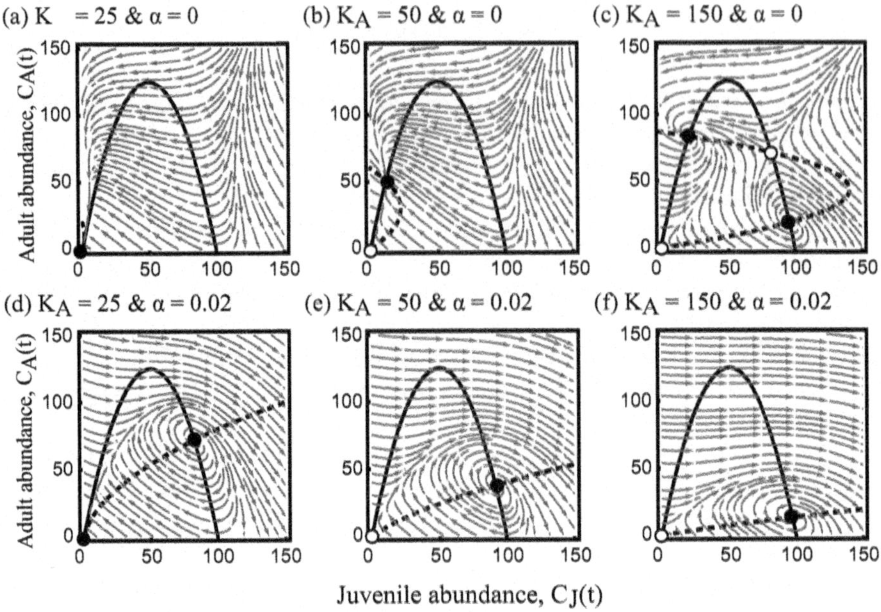

Fig. 1.2 The community dynamics of consumer-mediated ontogenetic antagonism–mutualism coupling, based on nullcline analysis when the adults are (**a–c**) antagonists ($a_A > \alpha$) and (*d–f*) mutualists ($\alpha > a_A$). The solid and dashed lines represent the nullclines (2a) and (2b), respectively, on the C_J^*–C_A^* plane. The solid and open circles represent stable and unstable equilibria, respectively. In the cases of antagonistic adults, the consumer (**a**) always goes extinct, (**b**) always persists, or (**c**) exhibits either a juvenile-dominated or an adult-dominated state, depending on the initial conditions. In the case of mutualistic adults, the consumer (**d**) either goes extinct or persists depending on the initial condition or (**e, f**) always persists. K_A = (**a, d**) 25, (**b, e**) 50, and (**c, f**) 150. α = (**a–c**) 0 and (**d–f**) 0.02. Other parameters are: $r_i = 1$, $K_J = 100$, $a_i = 0.01$, $b_i = 0.5$, and $d_i = 0.1$

capacities (K_J and K_A) are small, Eqs. (1.2a) and (1.2b) intersect only at the origin (Fig. 1.2a), which means that the consumer always goes extinct. When K_A becomes higher, Eq. (1.2b) shifts to the upper right, enabling the consumer to persist at a positive abundance (Fig. 1.2b). Intriguingly, a further increase in K_A generates alternative community states, where the dynamics converge to either the juvenile-dominated or the adult-dominated state, depending on the initial conditions (Fig. 1.2c). Increasing K_J shows a generally similar pattern; i.e., the consumer always goes extinct in unproductive environments, whereas the alternative community states occur in productive environments (see Schreiber and Rudolf 2008; Nakazawa 2011a, 2011b for details). The underlying mechanism for the occurrence of alternative community states is the positive feedback caused by two negative indirect effects between stage-specific resources. Suppose the juvenile resource increases. It promotes juvenile maturation and reduces the adult resource in a manner similar to apparent competition, which, in turn, suppresses adult reproduction and leads to a further increase in the juvenile resource in the same manner. In this scenario, the composition of the food web may drastically change following an environmental perturbation (regime shifts), and once the system is trapped in another stable state, environmental manipulation may not be effective for recovering the original state (hysteresis). Consequently, ontogenetic diet shifts can undermine the resilience of stage-structured food webs.

Next, consider that the adults are mutualists ($\alpha > a_A$) and Eq. (1.2a) is a left convex function (dashed curves in Figs. 1.2d–f). When both K_J and K_A are extremely low, Eqs. (1.2a) and (1.2b) intersect only at the origin, and thus, the consumer always goes extinct (not shown). As K_J increases, however, Eq. (1.2b) shifts downward, generating alternative community states with or without the consumer (Fig. 1.2d). In this situation, the consumer cannot invade the consumer-free equilibrium because the origin is locally stable. Nevertheless, the consumer can exist if the initial abundance is sufficiently high, i.e., Allee effect in mutualism (Holland and DeAngelis 2010). When K_A becomes higher, Eq. (1.2b) shifts downward. As a result, the consumer-free equilibrium becomes unstable and the consumer always coexists with the resource (Fig. 1.2e–f). Increasing K_J shows a generally similar pattern, i.e., alternative consumer-free and consumer-associated states occur in unproductive environments, whereas the dynamics converge to a single coexistence state in productive environment (see Ke and Nakazawa 2018 for details). The disappearance of alternative juvenile-dominated and adult-dominated states occurs because the positive feedback between stage-specific resources in the corresponding food-web module (see above) switches to negative when the adults are mutualists. As a result, if the juvenile resource increases, it promotes juvenile maturation and increases the adult resource, leading to apparent mutualism, which, in turn, facilitates adult reproduction and leads to a decrease in the juvenile resource.

Comparison of the two situations described above (i.e., when $a_A > \alpha$ and when $\alpha > a_A$) reveals that consumer-mediated ontogenetic antagonism–mutualism coupling is stabilizing. Specifically, it increases the potential for species coexistence in unproductive environments by generating an alternative consumer-associated state (compare Fig. 1.2a, d), while improving community resilience in productive environments by suppressing alternative juvenile-dominated and adult-dominated states (compare Fig. 1.2c, f). Overall, I suggest that consumer-mediated ontogenetic antagonism–mutualism coupling can play a crucial role in stabilizing stage-structured hybrid (e.g. plant–animal) communities under environmental changes. It should also be noted that the stabilizing effects are robust even when the interactions are nonlinear (see Ke and Nakazawa 2018 for details).

1.4 Resource-Mediated Ontogenetic Antagonism–Mutualism Coupling

Next, assume that only the resource species is stage structured, and it interacts with different consumer species at the juvenile and adult stages. For flowering plants, a common situation would be that the plant species is attacked by an antagonistic species at the juvenile (vegetative) stage, while there is a mutualistic relationship with another species at the adult (reproductive) stage (Fig. 1.1b). The community module for this scenario can be formulated as follows:

$$\frac{dR_J}{dt} = rR_A(1 - c_{JA}R_J - c_{AA}R_A) - mR_J(1 - c_{JJ}R_J - c_{AJ}R_A) \\ -d_{RJ}R_J - F_J(R_J)C_J + B(C_A)R_A \tag{1.3a}$$

$$\frac{dR_A}{dt} = mR_J(1 - c_{JJ}R_J - c_{AJ}R_A) - d_{RA}R_A - F_A(R_A)C_A \tag{1.3b}$$

$$\frac{dC_J}{dt} = [b_JF_J(R_J) - d_{CJ}]C_J \tag{1.3c}$$

$$\frac{dC_A}{dt} = [b_AF_A(R_A) - d_{CA}]C_A, \tag{1.3d}$$

where R_i ($i = J$ or A) is the abundance of juvenile or adult resource and C_i is the stage-specific consumer abundance. The resource exhibits density-dependent maturation (m) and reproduction (r) under stage-specific competition, where c_{ij} represents the strength of competition (the competition coefficient) from stage i to j. The parameter d_x ($x = RJ$, RA, CJ, or CA) is the stage- or species-specific death rate. The function $F_i(R_i)$ is the functional response of resource exploitation and b_i is its conversion efficiency. The mutualistic consumer C_A facilitates recruitment of the juvenile resource R_J according to the function $B(C_A)$. All the interaction terms are again assumed to be linear (the Holling's type-I), that is, $F_i(R_i) = a_iR_i$ and $B(C_A) = \alpha C_A$, where a_i ($i = J$ or A) is stage-specific resource exploitation (herbivory or nectar/pollen feeding) and α is the efficiency of mutualism at the adult stage.

Note that when $\alpha = 0$, the adult-specific consumer C_A does not contribute to the population growth of the resource, and thus, the consumer is considered to be antagonistic. In this situation, the community structure of model (3) is conceptually similar to that of the food-web module of de Roos et al. (2008), in which two different predators feed on the juvenile and adult stages of the same prey species, although the detailed formulation is different from mine. Hereafter, I compare the dynamics for situations when $\alpha = 0$ and when $\alpha > 0$ (i.e., when the adult-specific consumer is an antagonist and a mutualist, respectively), to illustrate the importance of ontogenetic antagonism–mutualism coupling (Fig. 1.3).

Importantly, alternative community states cannot occur in model (3) because there is only a single coexistence equilibrium, solved as follows:

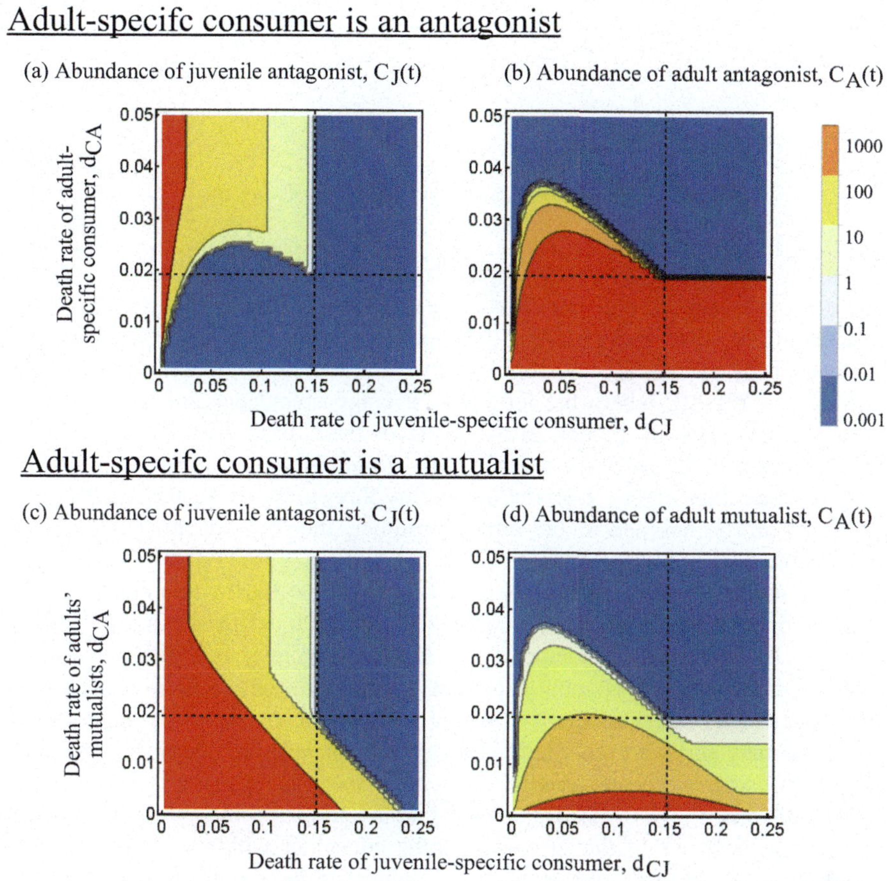

Fig. 1.3 The community dynamics of resource-mediated ontogenetic antagonism–mutualism coupling based on numerical simulations, when the adult-specific consumer is (**a**, **b**) an antagonist ($\alpha = 0$) and (**c**, **d**) a mutualist ($\alpha = 0.05$). The long-term average abundances of (**a**, **c**) the juvenile-specific consumer and (**b**, **d**) the adult-specific consumer are plotted in the plane of their death rates. Red and blue colors indicate high and low abundances, respectively. The vertical and horizontal dashed lines denote the invasibility thresholds above which the juvenile-specific and adult-specific consumers, respectively, cannot persist alone. If a consumer exists above the threshold, it illustrates the occurrence of apparent mutualism due to the other consumer. Parameters are $r = m = 1$, $d_{RJ} = d_{RA} = 0.1$, $c_{ii} = 0.02$, $c_{AJ} = 0.1$, $c_{JA} = 0.001$, $a_i = 0.01$, and $b_i = 0.5$

$$R_J^* = \frac{d_{CJ}}{b_J a_J} \tag{1.4a}$$

$$R_A^* = \frac{d_{CA}}{b_A a_A} \tag{1.4b}$$

$$C_A^* = \frac{1}{a_A}\left[\frac{mR_J^*}{R_A^*}(1 - c_{JJ}R_J^* - c_{AJ}R_A^*) - d_{RA}\right] \quad (1.4c)$$

$$C_J^* = \frac{1}{a_J}\left[\frac{rR_A^*}{R_J^*}(1 - c_{JA}R_J^* - c_{AA}R_A^*) + \frac{aC_A^*R_A^*}{R_J^*} - m(1 - c_{JJ}R_J^* - c_{AJ}R_A^*) - d_{RJ}\right]. \quad (1.4d)$$

This is true and analytically robust, irrespective of whether the adult-specific consumer is antagonistic or mutualistic, as long as all the species exist.

It should be noted, however, that the community dynamics can be more complex as revealed by numerical simulations below. The results of numerical analysis are provided in Fig.1.3, where the long-term average abundances of the two stage-specific consumers are plotted in the plane of their death rates. In addition, the maximum thresholds of death rates for their invasibility are also specified numerically (dashed lines in Fig. 1.3); the invisibility thresholds were calculated for each consumer in the absence of the other consumer, thereby indicating the species-specific potential for persistence.

If the adult-specific consumer is an antagonist ($\alpha = 0$), the two consumers are unlikely to coexist (see the small parameter space for co-occurrence in Fig. 1.3a, b). This can be explained by competitive exclusion principle, as the two consumers share the same resource. Notably, however, this is not the only case, especially when the death rate of either species is relatively high. For example, the adult-specific consumer may survive even when its death rate exceeds the invasibility threshold (Fig. 1.3b), where coexistence with the juvenile-specific consumer is possible (e.g., $d_{CJ} = 0.05$ and $d_{CA} = 0.03$). This means that the adult-specific consumer can exist only in the presence of the juvenile-specific consumer within a parameter space where it cannot persist alone otherwise. In other words, the juvenile-specific consumer may help the adult-specific consumer survive (i.e., apparent mutualism). This prediction is counterintuitive because the two consumers are potentially competitive when they share the same resource. de Roos et al. (2008) explained that the apparently counterintuitive outcome is due to overcompensation arising from stage structure. Suppose the juvenile-specific consumer increases in abundance. It reduces the abundance of the juvenile resource but relaxes the competition within the juvenile stage of the resource, which can promote the total maturation flux into the adult stage in the long run, potentially enabling the adult-specific consumer to persist (Fig. 1.3b). This is well known as thinning effect in plant biology. Mechanistically, this apparent facilitation can work in both directions, that is, the adult-specific consumer may also help in the persistence of the juvenile-specific consumer. As such, de Roos et al. (2008) suggested that "stage-specific predator species may help each other to persist while competing for a single prey." However, such mutual positive effects were rarely observed in my numerical analysis. This is because an asymmetric competition was assumed between stages (i.e., $c_{AJ} > c_{AA}$, $c_{JJ} > c_{JA}$),

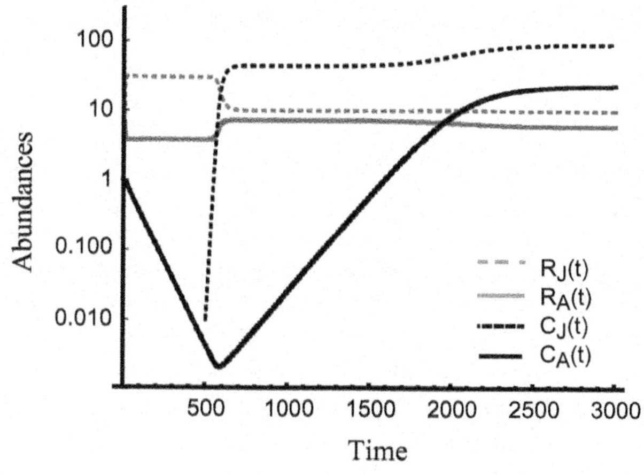

Fig. 1.4 An example of community dynamics illustrating the occurrence of apparent mutualism. The adult-specific consumer goes extinct without the juvenile-specific consumer but turns to recover once the juvenile-specific consumer is introduced at $t = 500$, after which the juvenile resource decreases but the adult resource increases. $d_{CJ} = 0.05$ and $d_{CA} = 0.03$. Other parameters are the same as in Fig.1.3

based on the common pattern in plants that adults are more competitive than juveniles.

Next, consider the case that the adult-specific consumer is a mutualist ($\alpha > 0$). Obviously, the adult-specific consumer would increase the abundance of the juvenile resource through mutualism and promote the juvenile-specific consumer. Furthermore, as shown above, the juvenile-specific consumer may also promote the persistence of the adult-specific consumer via the thinning effect, when its death rate d_{CA} is sufficiently high (e.g., $d_{CJ} = 0.05$ and $d_{CA} = 0.03$ in Figs. 1.3d). Ecologically, this prediction suggests, for example, that herbivores (i.e., antagonists of the vegetative stage of plants) may help pollinators (e.g., mutualists of the reproductive stage) persist (see Fig. 1.4 for their community dynamics). This prediction remains to be tested empirically. However, considering that the thinning effect is commonly observed in plants and thinning is in fact a common strategy in agriculture and forestry, such apparent facilitation between stage-specific antagonists and mutualists is likely to occur under some circumstances in reality. Here I do not analyze the present model in more detail, as apparent facilitation between stage-specific antagonists has already been studied in depth and is generally supported by a previous study (see de Roos et al. 2008 for more details).

Comparison of the two situations described above (i.e., when $\alpha = 0$ and when $\alpha > 0$) reveals that resource-mediated ontogenetic antagonism–mutualism coupling would be stabilizing. Specifically, it increases the potential for coexistence of stage-specific consumers (see wider parameter space of the co-occurrence in Fig. 1.3c, d than in Fig. 1.3a, b) as they do not have a purely competitive relationship, whereas the effects of apparent facilitation exist due to the stage structure of the shared resource. Further, this implies that plant-mediated ontogenetic coupling of herbivory and pollination could play a crucial role in the coexistence of herbivorous and pollinating insects and the maintenance of their diversity.

1.5 Conclusions and Future Perspectives

In this chapter, I briefly overviewed the dynamical features of two distinct modules for stage-structured hybrid communities, that is, consumer-mediated and resource-mediated ontogenetic antagonism–mutualism coupling. For both modules, ontogenetic antagonism–mutualism coupling is predicted to have stabilizing effects on the community structure, although the definition of stabilization is context dependent (i.e., higher community resilience and greater coexistence potential). One may be tempted to link those stabilizing effects to those found in previous nonstructured hybrid network models (Mougi and Kondoh 2012). However, while it would be intriguing to argue that interaction diversity is generally stabilizing irrespective of whether species are stage structured or not, caution should be exercised, first because of the differences in stability definitions, and second because of the differences in basic model structure. Here I evaluated community resilience or coexistence potential in simple module models, whereas previous studies have evaluated the local stability of coexistence equilibrium in network models (Mougi and Kondoh 2012). For a more detailed comparative discussion, stage-structured hybrid community models would need to be studied.

Here, I provide two perspectives that are important to establish the novel view of stage-structured hybrid communities. First, as mentioned above, it is absolutely necessary to study ontogenetic antagonism–mutualism coupling in larger network community models. In particular, bipartite network models with multiple stage-structured resources and consumers would be desired (Fig. 1.5), considering the ubiquity of ontogenetic herbivory–pollination coupling in plant–insect interactions

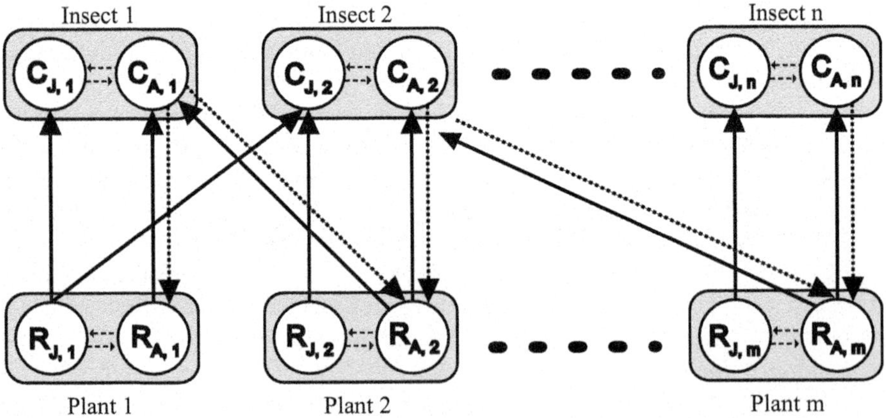

Fig. 1.5 Conceptual diagram of a stage-structured bipartite network model, with m resources and n consumers (adapted from Ke and Nakazawa 2018). Gray boxes represent the species identity in which maturation and reproduction couple juvenile and adult stages. Interspecific interactions denoted by solid and dashed lines indicate resource exploitation (e.g., herbivory and pollen/nectar feeding) and mutualism (e.g., pollination), respectively. Note that not all links are shown for simplicity

and their great biodiversity. Such a stage-structured bipartite system can be formulated as follows:

$$\begin{aligned}\frac{dR_{J,i}}{dt} &= r_i R_{A,i}\left(1-\sum_{x=1}^{m} c_{J_x A_i} R_{J,x} - \sum_{x=1}^{m} c_{A_x A_i} R_{A,x}\right)\\ &-m_i R_{J,i}\left(1-\sum_{x=1}^{m} c_{J_x J_i} R_{J,x} - \sum_{x=1}^{m} c_{A_x J_i} R_{A,x}\right)\\ &-d_{RJ,i} R_{J,i} - \sum_{x=1}^{n} F_{J,x,i}(R_{J,i}) C_{J,x} + \sum_{x=1}^{n} B_{x,i}(C_{A,x}) R_{A,i}\end{aligned} \tag{1.5a}$$

$$\begin{aligned}\frac{dR_{A,i}}{dt} &= m_i R_{J,i}\left(1-\sum_{x=1}^{m} c_{J_x J_i} R_{J,x} - \sum_{x=1}^{m} c_{A_x J_i} R_{A,x}\right)\\ &-d_{RA,i} R_{A,i} - \sum_{x=1}^{n} F_{A,x,i}(R_{A,i}) C_{A,x}\end{aligned} \tag{1.5b}$$

$$\frac{dC_{J,i}}{dt} = \sum_{x=1}^{m} b_{A,i,x} F_{A,i,x}(R_{A,x}) C_{A,i} - \sum_{x=1}^{m} b_{J,i,x} F_{J,i,x}(R_{J,x}) C_{J,i} - d_{CJ,i} C_{J,i} \tag{1.5c}$$

$$\frac{dC_{A,i}}{dt} = \sum_{x=1}^{m} b_{J,i,x} F_{J,i,x}(R_{J,x}) C_{J,i} - d_{CA,i} C_{A,i}, \tag{1.5d}$$

where m species of resources (plants) and n species of consumers (insects) are all assumed to have a juvenile–adult (vegetative–reproductive) stage structure; the detailed definitions of parameters and functions are provided above and thus omitted here. Note that purely herbivorous insects (e.g., grasshoppers) can be described by neutralizing mutualistic terms (i.e., $\alpha = 0$). While previous studies have predicted that the antagonism-to-mutualism ratio (Mougi and Kondoh) or connectance (Sauve et al. 2014) is a key for species coexistence in nonstructured hybrid networks, effects of these factors in stage-structured hybrid networks remain unknown. To analyze the effects, the model structure needs to be manipulated in a stage-specific manner using this model.

Next, I emphasize that the view of stage-structured hybrid community should be applied not only for plant–animal interactions but also for other species relationships that have traditionally been considered as mutualistic systems. Mutualism creates both benefits and costs to organisms (i.e., consumer–resource approach to mutualism, Holland and DeAngelis 2010); for example, pollen/nectar consumption can be costly for plants in pollination mutualism. Importantly, the relative costs and benefits of mutualism vary quantitatively depending on the body size or life-history stage of the interacting organisms. Consequently, the interaction type may change ontogenetically between antagonism and mutualism if the stage or size structure is explicitly considered in mutualism (Nakazawa 2020). This idea is close to the antagonism–mutualism continuum, which emphasizes that antagonism and mutualism are not binary alternatives but context dependent (Bronstein 2015, Johnstone and Bshary 2002, Holland and DeAngelis 2009, 2010). However, the point here is that ontogenetic niche shifts play an important role in linking antagonism and mutualism along the continuum in hybrid communities. Recall that the ontogenetic growth in body size is the most fundamental biological aspect and commonly observed in multicellular organisms (Wilbur 1980; Werner and Gilliam 1984). Direct and indirect evidence have suggested that apparently mutualistic relationships often involve ontogenetic antagonism–mutualism coupling. Examples include seed dispersal mutualism between plants and birds (Holland et al. 2006), nutritional mutualism between plants and mycorrhizal fungi (Johnson et al. 1997), defensive mutualism between aphids and ants (Breton and Addicott 1992; Stadler and Dixon 2005), and cleaning mutualism between hosts and cleaner fish (Grutter 2000; Cole 2010; Skelton et al. 2014). In these examples, some stage of the individuals may act antagonistically, while other stages may exhibit mutualism. The community consequences of such ontogenetic niche shift in mutualism remain understudied but are expected to be critical and ubiquitous in nature (Nakazawa, in preparation).

I end this chapter by concluding that incorporating ontogenetic niche shifts in hybrid community dynamics (i.e., the stage-structured hybrid community view) will open a new research avenue for community ecology and biodiversity research. Future research efforts are warranted to advance our understanding of how ecological communities are modulated by stage structure and interaction diversity, using both theoretical and empirical approaches.

Acknowledgment TN was supported by the Ministry of Science and Technology of Taiwan.

References

Allesina S, Tang S (2012) Stability criteria for complex ecosystems. Nature 483:205–208

Allesina S, Tang S (2015) The stability–complexity relationship at age 40: a random matrix perspective. Popul Ecol 57:63–75

Altermatt F, Pearse IS (2011) Similarity and specialization of the larval versus adult diet of European butterflies and moths. Am Nat 178:372–382

Bascompte J, Jordano P (2014) Mutualistic networks. Princeton University Press, Princeton

Begon M, Townsend CR, Harper JL (2006) Ecology: from individuals to ecosystems, 4th edn. Blackwell, Oxford

BoegeK, BartonKE, DirzoR (2011) Influence of tree ontogeny on plant–herbivore interactions. In: Meinzer FC, LachenbruchB, DawsonTE (eds)Size- and age-related changes in tree structure and function. Springer,Cham, pp 132–214

Breton LM, Addicott JF (1992) Density-dependent mutualism in an aphid–ant interaction. Ecology 73:2175–2180

Bronstein JL (2015) Mutualism. Oxford University Press, Oxford

Bronstein JL, Huxman TE, Davidowitz G (2007) Plant-mediated effects linking herbivory and pollination. In: Ohgushi T, Craig TP, Price PW (eds) Ecological communities: plant mediation in indirect interaction webs. Cambridge University Press, Cambridge, pp 75–103

Cole AJ (2010) Cleaning to corallivory: ontogenetic shifts in feeding ecology of tubelip wrasse. Coral Reefs 29:125–129 de Roos AM, Persson L (2013) Population and community ecology of ontogenetic development. Princeton University Press, Princeton

de Roos AM, Schellekens T, Van Kooten T, Persson L (2008) Stage-specific predator species help each other to persist while competing for a single prey. Proc Natl Acad Sci U S A 105:13930–13935

Grutter AS (2000) Ontogenetic variation in the diet of the cleaner fish *Labroides dimidiatus* and its ecological consequences. Mar Ecol Progr Ser 197:241–246

Guill C (2009) Alternative dynamical states in stage-structured consumer populations. Theor Popul Biol 76:168–178

Holland JN, DeAngelis DL (2009) Consumer–resource theory predicts dynamic transitions between outcomes of interspecific interactions. Ecol Lett 12:1357–1366

Holland JN, DeAngelis DL (2010) A consumer–resource approach to the density-dependent population dynamics of mutualism. Ecology 91:1286–1295

Holland JM, Hutchinson MAS, Aebischer NJ (2006) A review of invertebrates and seed-bearing plants as food for farmland birds in Europe. Ann Appl Biol 148:49–71

Ikemoto M, Ida TY, Utsumi S, Ohgushi T (2017) Community-wide impacts of early season herbivory on flower visitors on tall goldenrod. Ecol Entomol 42:164–172

Ives AR, Carpenter SR (2007) Stability and diversity of ecosystems. Science 317:58–62

Johnson NC, Graham JH, Smith FA (1997) Functioning of mycorrhizal associations along the mutualism–parasitism continuum. New Phytol 135:575–585

Johnstone RA, Bshary R (2002) From parasitism to mutualism: partner control in asymmetric interactions. Ecol Lett 5:634–639

Ke PJ, Nakazawa T (2018) Ontogenetic antagonism–mutualism coupling: perspectives on resilience of stage-structured communities. Oikos 127:353–363

Kondoh M, Mougi A (2015) Interaction-type diversity hypothesis and interaction strength: the condition for the positive complexity–stability effect to arise. Popul Ecol 57:21–27

Lurgi M, Montoya D, Montoya JM (2016) The effects of space and diversity of interaction types on the stability of complex ecological networks. Theor Ecol 9:3–13

May RM (1972) Will a large complex system be stable? Nature 238:413–414

McCann KS (2011) Food webs. Princeton University Press, Princeton

Mcnaughton SJ (1978) Stability and diversity of ecological communities. Nature 274:251–253

Miller TEX, Rudolf VHW (2011) Thinking inside the box: community-level consequences of stage-structured populations. Trend Ecol Evol 26:457–466

Mougi A (2017) Persistence of a stage-structured food-web. Sci Rep 7:11055

Mougi A, Kondoh M (2012) Diversity of interaction types and ecological community stability. Science 337:349–351

Nakazawa T (2011a) Ontogenetic niche shift, food-web coupling, and alternative stable states. Theor Ecol 4:479–494

Nakazawa T (2011b) Alternative stable states generated by ontogenetic niche shift in the presence of multiple resource use. PLoS One 6:e14667

Nakazawa T (2014) A dynamic resilience perspective toward integrated ecosystem management: biodiversity, landscape, and climate. In: Okuda N (ed) Springer briefs in biology: biodiversity in aquatic systems and environments: Lake Biwa. Springer, Tokyo, pp 69–91

Nakazawa T (2015) Ontogenetic niche shifts matter in community ecology: a review and future perspectives. Popul Ecol 57:347–354

Nakazawa T (2020) A perspective on stage-structured mutualism and its community consequences. Oikos 129:297–310

Pimm SL (1984) The complexity and stability of ecosystems. Nature 307:321–326

Sauve AMC, Fontaine C, Thébault E (2014) Structure–stability relationships in networks combining mutualistic and antagonistic interactions. Oikos 123:378–384

Schreiber S, Rudolf VHW (2008) Crossing habitat boundaries: coupling dynamics of ecosystems through complex life cycles. Ecol Lett 11:576–587

Skelton J, Creed RP, Brown BL (2014) Ontogenetic shift in host tolerance controls initiation of a cleaning symbiosis. Oikos 123:677–686

Stadler B, Dixon AFG (2005) Ecology and evolution of aphid–ant interactions. Ann Rev Ecol Evol Syst 36:345–372

Verhoef HA, Morin PJ (2010) Community ecology: processes, models and applications. Oxford University Press, Oxford

Wäckers FL, Romeis J, van Rijn P (2007) Nectar and pollen feeding by insect herbivores and implications for multitrophic interactions. Annu Rev EntomolAnn Rev Entomol 52:301–323

Werner EE, Gilliam JF (1984) The ontogenetic niche and species interactions in size-structured populations. Ann Rev Ecol Syst 15:393–425

Wilbur HM (1980) Complex life cycles. Ann Rev Ecol Syst 11:67–93

Chapter 2
Effects of Rapid Evolution on Population Cycles and Extinction in Predator–Prey Systems

Masato Yamamichi

Abstract Traditionally ecologists have tended to consider species as the smallest unit for understanding community dynamics. However, recent studies demonstrated that there is substantial amount of intraspecific genetic variation and the resultant microevolution can affect ecological dynamics. Here I introduce recent studies showing how adaptive phenotypic changes affect community dynamics, especially in predator–prey systems. First, I review the effects of rapid evolution on predator–prey cycles. Experimental and theoretical studies have demonstrated that the phase lag between predator and prey densities can be changed from the classic quarter to the half ("antiphase cycles") or three-quarters ("clockwise cycles") by prey defense evolution or coevolution, respectively. In addition, prey defense evolution can cause "cryptic cycles" where predator densities fluctuate whereas prey densities stay almost constant. Second, I explain how rapid adaptive evolution can prevent extinction ("evolutionary rescue"). Evolutionary rescue with interspecific interactions can create counterintuitive dynamics and will be important for understanding species coexistence in communities. Finally, I discuss future perspectives of empirical and theoretical studies on eco-evolutionary dynamics in complex communities.

Keywords Adaptive evolution · Coevolution · Eco-evolutionary dynamics · Evolutionary rescue · Phenotypic plasticity · Predator–prey cycles · Rapid contemporary evolution · Species coexistence

> "Density influences selection; selection influences genetic make-up; and in turn, genetic make-up influences density. The actions and reactions of the interacting populations in the food chain cycling in this mechanism result in the evolution and regulation of animal populations." Pimentel (1961)

M. Yamamichi
School of Biological Sciences, The University of Queensland, Brisbane, QLD, Australia
e-mail: m.yamamichi@uq.edu.au

A. Mougi (ed.), *Diversity of Functional Traits and Interactions*, Theoretical Biology,
https://doi.org/10.1007/978-981-15-7953-0_2

2.1 Introduction

Ecologists have tended to consider species as the smallest unit, and tried to understand population and community dynamics from species abundances (Lotka 1925; Volterra 1926; Rosenzweig and MacArthur 1963; May 1972; Tilman 1982; Chesson 2000). On the other hand, there has been an increasing interest on ecological consequences of intraspecific trait variation due to age- or stage-structures (Chap. 1 of this book), sexual dimorphism (Chap. 3 of this book), phenotypic plasticity, and genetic variation. This view suggests that not only the abundance (i.e., quantity) of species, but also the phenotypic variation (i.e., quality) within species are important in community ecology (Bolnick et al. 2011; Des Roches et al. 2018).

In this chapter, I focus on ecological consequences of intraspecific genetic variation and adaptive trait changes. While classic studies tended to assume that evolutionary processes are much slower than ecological processes (Darwin 1859; Slobodkin 1961), recent studies have shown that the time-scale separation between them is actually artificial: there is substantial amount of intraspecific genetic variation for selection to work on, and selection pressure is often fluctuating temporally and spatially (Hoekstra et al. 2001; Kingsolver et al. 2001; Bell 2010), thereby creating rapid contemporary evolution in short time scales (Gingerich 1993; Hendry and Kinnison 1999; Messer et al. 2016). This is a kind of microevolution, where allele frequencies change through time in populations (Box 2.1). The speed of evolution can be sometimes rapid enough to affect contemporary ecological processes (Hairston et al. 2005) as adaptive behavior (Abrams 2010). Because ecological processes can alter evolutionary dynamics by changing fitness landscapes, there may be feedbacks between ecological and evolutionary processes (eco-evolutionary feedbacks: Thompson 1998; Hairston et al. 2005; Fussmann et al. 2007; Johnson and Stinchcombe 2007; Kinnison and Hairston 2007; Schoener 2011; Ellner 2013; Reznick 2013; Hendry 2016). The novel findings suggest that we need to re-consider classic theory of community ecology from a new perspective of eco-evolutionary dynamics.

Eco-evolutionary dynamics is a rapidly growing field in ecology and evolutionary biology: the number of papers with "eco-evo" has increased significantly in the last two decades (Fig. 2.1a) and there have been many special issues of ecological journals featuring eco-evolutionary dynamics (Table 2.1). Actually, the idea of feedbacks between ecological and evolutionary processes was already proposed in the 1960s (Chitty 1960; Pimentel 1961). It was called "genetic feedback" by Pimentel (1961), and the researchers were mostly interested in the effects of evolution on population regulation (i.e., whether rapid evolution stabilizes or destabilizes population dynamics). The idea needed to wait for a few decades before becoming popular until empirical evidence of rapid evolution accumulated since the 1990s. This resulted in interesting citation patterns of the classic papers by David Pimentel: after the long declining trend from the 1970s to the 1990s, the number of citations increased since ca. 2000 (Fig. 2.1b). Currently researchers are investigating the effects of rapid evolution not only on population dynamics, but also on community

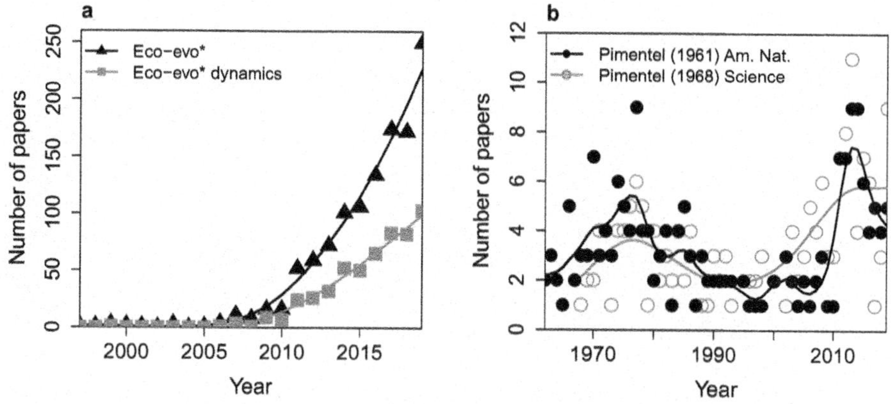

Fig. 2.1 (**a**) Papers with "eco-evo*" (black triangles: $n = 1187$) and "eco-evo* dynamics" (gray boxes: $n = 540$) are shown. The search was conducted on June 29, 2020 by using ISI Web of Science (after Shefferson and Salguero-Gómez 2015; Govaert 2018). (**b**) Papers citing Pimentel (1961) (black circles: $n = 185$) and Pimentel (1968) (gray circles: $n = 170$). Lines are spline smooths of the data by the "`smooth.spline`" function of R (R Core Team 2019)

Table 2.1 Special issues on eco-evolutionary dynamics

Year	Journal	Theme
2003	*Ecology*	Community genetics: new insights into community ecology by integrating population genetics
2007	*Ecol Res*	Evolution in biological invasion
2007	*Funct Ecol*	Causes and consequences of adaptive evolution
2009	*Phil Trans R Soc B*	Eco-evolutionary dynamics
2011	*Phil Trans R Soc B*	Community genetics: at the crossroads of ecology and evolutionary genetics
2013	*Phil Trans R Soc B*	Evolutionary rescue in changing environments
2013	*Am Nat*	A critical look at reciprocity in ecology and evolution
2014	*Adv Ecol Res*	Eco-evolutionary dynamics
2015	*J Ecol*	Eco-evolutionary dynamics in plants: interactive processes at overlapping time scales and their implications
2016	*Funct Ecol*	Ecosystems, evolution and plant–soil feedbacks
2017	*Copeia*	Eco-evolutionary dynamics in cold blood
2019	*Funct Ecol, J Anim Ecol* (joint special feature)	Eco-evolutionary dynamics across scales

(Johnson and Stinchcombe 2007) and ecosystem (Fussmann et al. 2007; Matthews et al. 2011) dynamics, and some of these studies have been called "community genetics" (Antonovics 1992) and "ecosystem genetics" (Whitham et al. 2006).

Here I focus on eco-evolutionary dynamics in a predator–prey system, which is one of the most fundamental and well-studied community modules. First, I introduce empirical and theoretical studies on the effects of rapid evolution on predator–prey cycles (reviewed in Fussmann et al. 2005, Jones et al. 2009b, Ellner 2013, Koch et al. 2014). Experimental and theoretical studies have demonstrated that the phase lag between predator and prey densities can be changed from the classic quarter to the half (i.e., the antiphase cycles) or three-quarter (i.e., the clockwise cycles) by prey defense evolution or coevolution, respectively. Additionally, predator densities can fluctuate while prey densities are almost constant when prey defense is not costly (i.e., cryptic cycles). This suggests that it may be difficult to understand complex ecological dynamics without considering rapid evolution. Then I explain how adaptation can prevent extinction by evolutionary rescue (Kinnison and Hairston 2007; Alexander et al. 2014; Carlson et al. 2014; Bell 2017) with interspecific interactions. If adaptive evolution occurs faster than population decline in deteriorated environments, rapid evolution may prevent extinction. While most previous studies have investigated evolutionary rescue in the context of population ecology (i.e., single-species dynamics), recent studies revealed that interspecific interactions can create complex dynamics such as indirect evolutionary rescue where prey evolution prevents predator extinction (Yamamichi and Miner 2015; Cortez and Yamamichi 2019), as well as evolutionary suicide in which adaptive evolution results in extinction (Webb 2003; Parvinen 2005). This will be important for theory dealing with species coexistence in complex communities with rapid evolution. Finally, I discuss future perspectives of community ecology with rapid evolution, focusing on (1) the mechanisms of adaptive phenotypic changes, (2) types of ecological interactions, (3) scales of communities, and (4) eco-evolutionary dynamics in the wild.

Box 2.1: How Does Fluctuating Selection Maintain Genetic Variation and Promote Evolution?

For adaptive evolution to occur, three conditions need to be satisfied: the presence of trait variation in a population, covariance between the trait and fitness, and heritability in the trait. While spatially fluctuating selection and negative frequency-dependent selection can effectively maintain genetic variation (Hedrick 2006), classic studies in population genetics revealed that it is difficult to maintain genetic variation under temporally fluctuating selection. This is because an allele with the highest geometric mean fitness eliminates other alleles in the simple Wright–Fisher model with non-overlapping generation and haploid inheritance (Dempster 1955; Haldane and Jayakar 1963). Despite this, temporally fluctuating selection, combined with buffering factors that limit the influence of selection on allele frequency dynamics, has the potential to maintain genetic variation and bring about rapid evolution.

Genomic studies on fruit fly (*Drosophila melanogaster*) found that many SNP polymorphisms are maintained with seasonal fluctuations (Bergland et al. 2014), suggesting the potential importance of temporally fluctuating selection. Theoretical studies demonstrated that the maintenance of genetic variation is

(continued)

Box 2.1 (continued)
possible by the storage effect (Chesson 2000) with the aid of various buffering factors including complete allele dominance (Haldane and Jayakar 1963), overlapping generations (Ellner and Hairston 1994; Yamamichi et al. 2019a), maternal effects (Yamamichi and Hoso 2017), epistasis (Gulisija et al. 2016), sex-limited traits (Reinhold 2000), and temporal changes of allele dominance (Wittmann et al. 2017; Bertram and Masel 2019).

Furthermore, recent studies revealed that the buffering factors have a double-edged effect for rapid evolution: a buffering factor (e.g., generation overlap) promotes the maintenance of genetic variation in temporally fluctuating selection (Fig. 2.2). At the same time, the buffering factor blurs the trait-fitness covariance and results in slowing down the speed of evolution. For example, a model with low generation overlap sharply responds to environmental fluctuations, whereas that with high generation overlap does not in Fig. 2.2. Thus, the intermediate buffering can maximize the speed of evolution and may drive eco-evolutionary dynamics (Yamamichi et al. 2019a). As in the maintenance of species diversity, fluctuating environments maintain genetic variation via the storage effect and relative nonlinearity (Chesson 2000). However, it is still unclear how relative nonlinearity and the storage effect simultaneously promote the maintenance of genetic variation and rapid evolution.

2.2 Eco-Evolutionary Dynamics in Communities

For ecologists to incorporate rapid contemporary evolution as an additional factor affecting community dynamics, there are two strategies: one is starting from the smallest module of community dynamics (i.e., two species systems) with detailed mechanistic understandings, and the other one is beginning from the largest scale (e.g., food webs and metacommunities) in the hope of finding consistent macro-scale patterns (Lawton 1999; Toju et al. 2017). In the following, I introduce a series of experimental and theoretical studies that employed the former approach for studying the effects of rapid evolution on predator–prey dynamics by utilizing plankton chemostats (continuous flow-through systems) to observe long-term community processes.

2.2.1 *Predator–Prey Cycles in Experiments*

Ecology studies how interspecific interactions and environments affect the abundance of organisms in nature. Interspecific interactions between exploiters (e.g., predators, herbivores, or parasites) and victims (e.g., prey, plants, or hosts) can drive

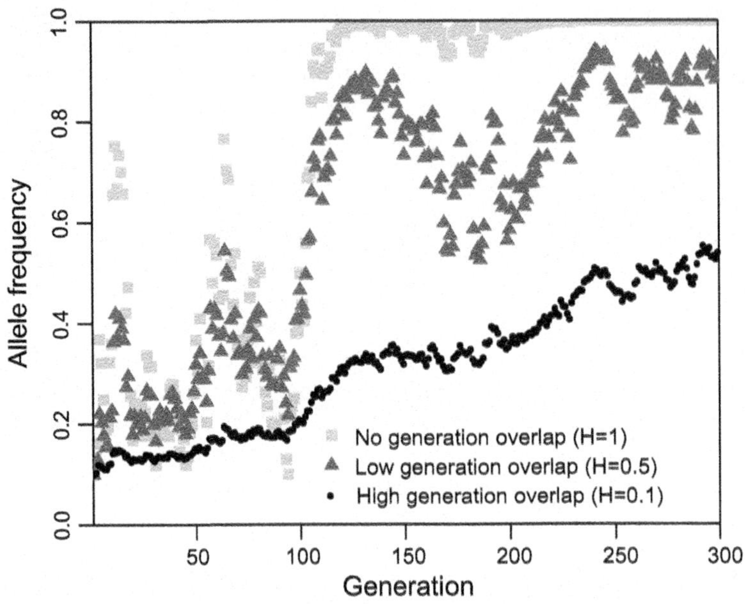

Fig. 2.2 Effects of generation overlap on allele frequency dynamics under temporally fluctuating selection. Simulation results of the Wright–Fisher model of haploid inheritance with different hatching rates from dormant propagules (H) are shown. Light-gray boxes, gray triangles, and black circles represent dynamics with no generation overlap ($H = 1$), low generation overlap ($H = 0.5$), and high generation overlap ($H = 0.1$), respectively. See Appendix for details of models and parameters

population cycles (Lotka 1925; Volterra 1926), and indeed, one review suggests that one-third of population exhibit periodic dynamics (Kendall et al. 1998). Thus, population and community ecologists have investigated how exploitation drives population cycles for a long time, and found that the phase lag between predator and prey densities is a quarter of a period (Figs. 2.3a, b and 2.4a, b, Box 2.2: Rosenzweig and MacArthur 1963, Bulmer 1975).

When researchers tried to explore chaotic cycles (Ellner 2013) by using chemostats with rotifers (*Brachionus calyciflorus*) and green algae (*Chlorella vulgaris*), they found strange cycles (Fig. 2.3c, d): while they succeeded in predicting the qualitative patterns (i.e., population cycles, equilibrium, and extinction) by model simulations (Fussmann et al. 2000), the empirical phase lengths were longer than the simulations and the phase lag was not consistent with the theoretical predictions (Fig. 2.3c, d). By fitting four models incorporating quality changes in addition to quantity changes to experimental results, it turned out that the best-fit model for explaining the observed strange patterns was that with the frequency change of prey genotypes (i.e., microevolution) (Shertzer et al. 2002). Indeed, a subsequent experimental study confirmed that the presence of genetic variation promotes the antiphase cycles, whereas the absence of genetic variation results in

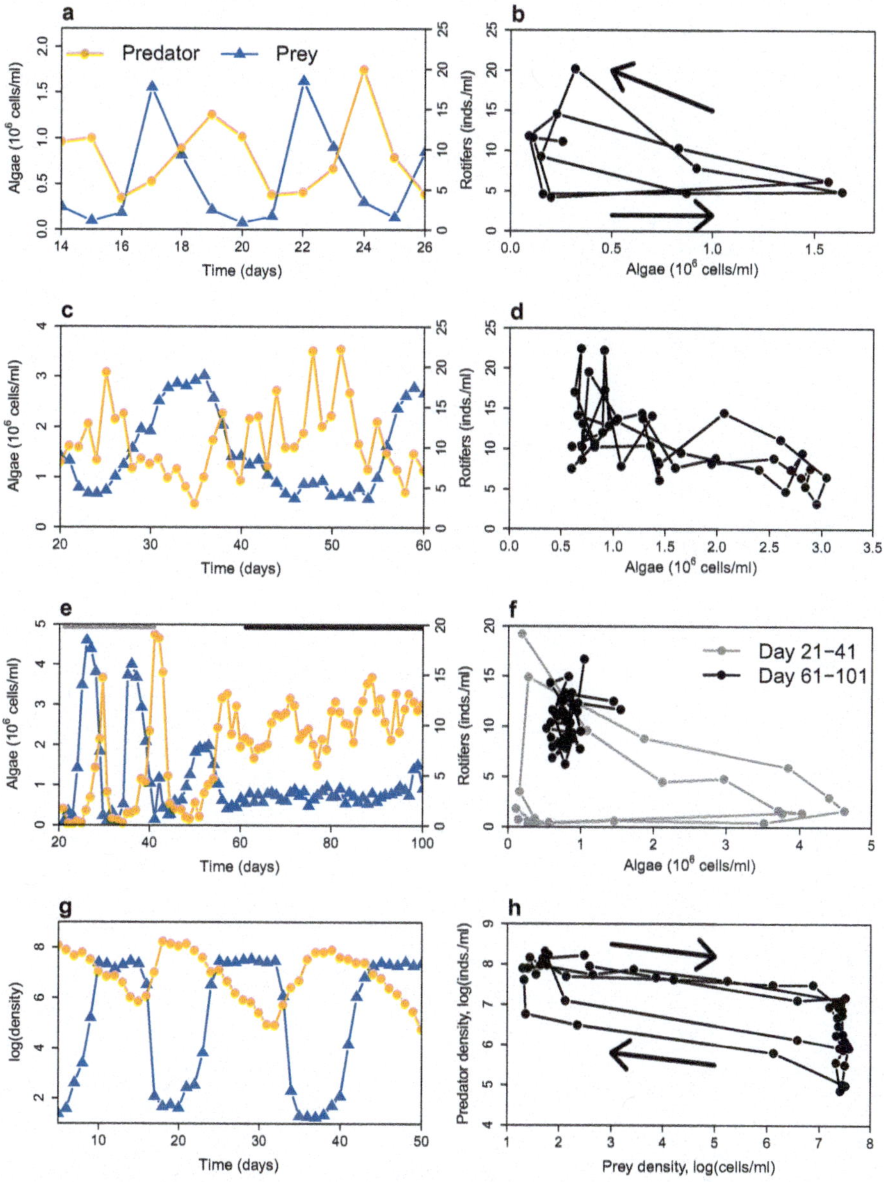

Fig. 2.3 Time-series data of classic (counterclockwise) cycles (**a**, **b**), antiphase cycles (**c**, **d**), cryptic cycles (**e**, **f**), and clockwise cycles (**g**, **h**). (**a**, **c**, **e**, **g**): Blue triangles represent victim species (green algae *Chlorella vulgaris* in **a**, **c**, and **e**, and blue-green algae *Plectonema boryanum* in **g**). Orange circles represent exploiter species (herbivorous rotifers *Brachionus calyciflorus* in **a**, **c**, and **e**, and LPP-I cyanophage in **g**). Note that the gray line is from day 21 to 41, and the black line is from day 61 to 101 in **f**. (**a–d**): Yoshida et al. (2003), (**e**, **f**): Yoshida et al. (2007), and (**g**, **h**): Cannon et al. (1976) based on Cortez (2018)

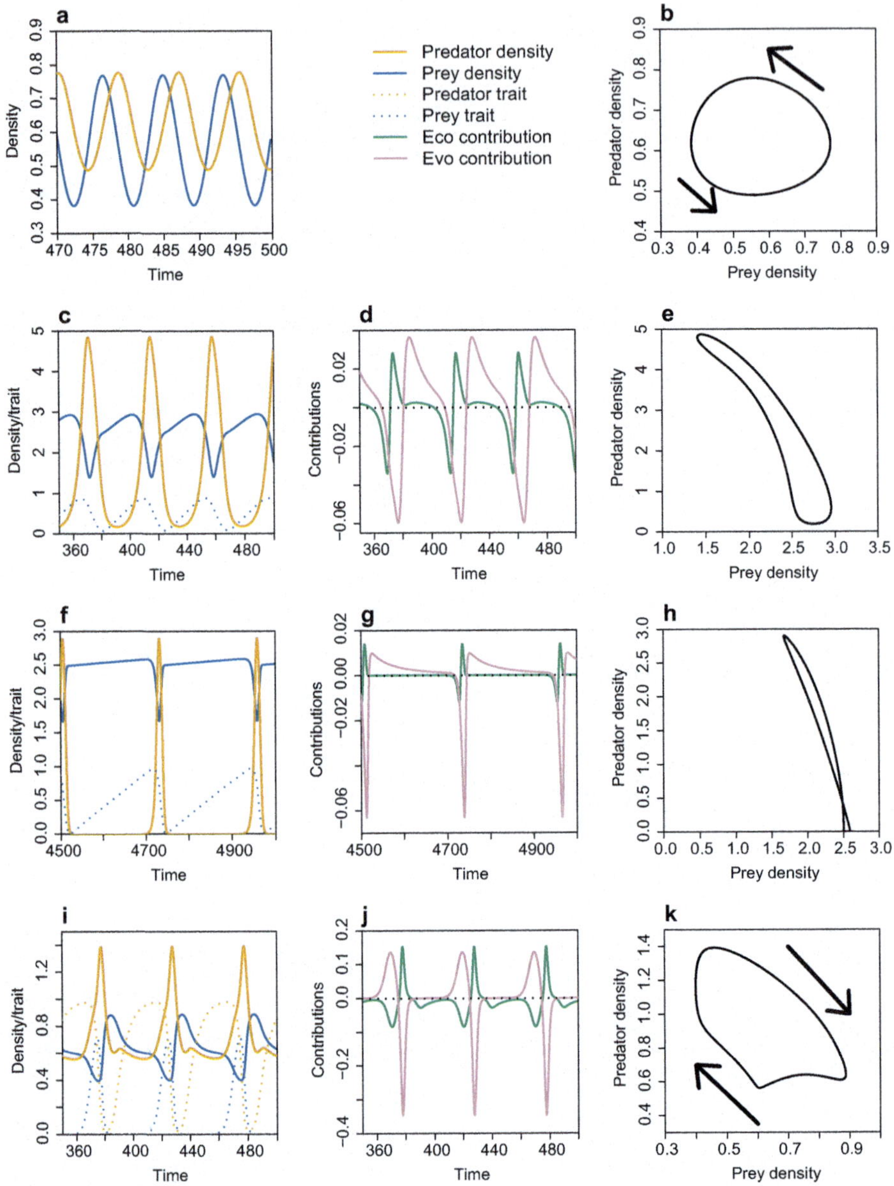

Fig. 2.4 Model simulations of classic (counterclockwise) cycles (**a**, **b**), antiphase cycles (**c**–**e**), cryptic cycles (**f**–**h**), and clockwise cycles (**i**–**k**). (**a**, **c**, **f**, **i**): Orange and blue lines represent predator and prey densities, respectively. Dotted orange and blue lines are predator and prey traits, respectively. (**d**, **g**, **j**): Green and magenta lines show contributions of prey density and prey trait to predator per-capita growth rate, respectively. Evo/eco ratio is 2.94 in antiphase cycles, 8.24 in cryptic cycles, and 1.74 in clockwise cycles. Models of **a**–**h** are based on Abrams and Matsuda (1997), and a model of **i**–**k** is from Cortez and Weitz (2014) (see Appendix)

the classic cycles with the quarter-period lag (Yoshida et al. 2003). Interestingly, the antiphase cycles driven by prey evolution were predicted by a theoretical model (Abrams and Matsuda 1997) three years before the observation (Fussmann et al. 2000). Rapid evolution in *Chlorella* was further confirmed by tracking the frequency change of genotypes using allele-specific quantitative PCR (Meyer et al. 2006). While it is difficult and time consuming to differentiate the defended and undefended genotypes of *Chlorella*, it is easy in *Chlamydomonas reinhardtii*, as they evolve to be defended by clumping in the presence of rotifers: the clumped genotypes are too large for rotifers to eat (Becks et al. 2010, 2012). The antiphase cycles were also observed in the rotifer–*Chlamydomonas* system as in the rotifer–*Chlorella* system. Furthermore, Hiltunen et al. (2014b) demonstrated that the antiphase cycles were observed not only in the plankton chemostats, but also in previous studies including bacteria experiments in the 1970s by using the EDI approach (see Box 2.2). Although other processes such as age structures can produce antiphase cycles (e.g., de Roos and Persson 2013; Hiltunen et al. 2014b), Hiltunen et al. (2014b) found that larger population size of prey species and longer experimental duration promote antiphase cycles, indicating the role of rapid evolution for driving the observed patterns. This suggests the prevalence of rapid defense evolution and resultant antiphase cycles.

In addition to antiphase cycles, Yoshida et al. (2007) found cryptic cycles, where prey densities stay almost constant, while predator densities fluctuate (Fig. 2.3e, f). There seems no interspecific interaction between rotifers and algae from the time-series data, but theory suggests that rapidly changing genetic composition in prey species can drive cryptic cycles when defense is not costly. Similar dynamics was found in phage–bacterium systems (Bohannan and Lenski 1999), indicating the potential importance of cryptic cycles in diverse systems.

2.2.2 *Predator–Prey Cycles in Models*

For understanding predator–prey cycles with rapid evolution, researchers have often considered a predator–prey model where densities of defended prey (x_1), undefended prey (x_2), and predator (y) are represented by

$$\begin{aligned} \frac{dx_i}{dt} &= x_i[\, f_i(x_1, x_2) - g_i(x_1, x_2, y)], \quad i = 1, 2, \\ \frac{dy}{dt} &= y[h(x_1, x_2, y) - d(y)], \end{aligned} \tag{2.1}$$

where f_i is prey reproduction, g_i is predation, h is predator reproduction, and d is mortality (Jones and Ellner 2004, 2007; Yoshida et al. 2007; Yamamichi et al. 2014). Undefended prey suffers more from predation ($g_1 < g_2$), whereas the growth rate of defended prey in the absence of predator is smaller due to a trade-off ($f_1 < f_2$). Because of direct resource competition and apparent competition, this diamond food web can result in coexistence of the two prey types (Kretzschmar et al. 1993; Holt et al. 1994; Abrams 1999).

The model with two prey genotypes (Eq. 2.1) can be approximated by a quantitative trait model (Ellner and Becks 2011; Cortez and Weitz 2014),

$$\begin{aligned}\frac{dx}{dt} &= x[f(x,\alpha,\overline{\alpha}) - g(x,y,\alpha,\overline{\alpha})]|_{\alpha=\overline{\alpha}},\\ \frac{dy}{dt} &= y[h(x,y,\overline{\alpha}) - d(y)],\\ \frac{d\overline{\alpha}}{dt} &= V\frac{\partial}{\partial\alpha}[f(x,\alpha,\overline{\alpha}) - g(x,y,\alpha,\overline{\alpha})]\Big|_{\alpha=\overline{\alpha}},\end{aligned} \tag{2.2}$$

where α is the prey quantitative trait, $\overline{\alpha}$ is the population mean of the prey quantitative trait, and V is additive genetic variance of the prey trait. Here the speed and direction of trait evolution are determined by additive genetic variance and local fitness gradient (Abrams 2001).

Because green algae in chemostats do not reproduce sexually, Jones and Ellner (2004, 2007) used the clonal model with defended and undefended prey genotypes (Eq. 2.1) for understanding the condition for antiphase cycles (Fig. 2.8). The clonal model itself may be more suitable for studying bistability (i.e., alternative stable states) and the effects of introduction timing of genetic variation on antiphase cycles (Yamamichi et al. 2014), whereas the quantitative trait model (Eq. 2.2) may be suitable for understanding the effects of genetic variation on eco-evolutionary dynamics (Cortez 2016, 2018; Cortez and Patel 2017).

In addition to the mechanistic model analyses, Hairston et al. (2005) developed a partition method called "Geber method" for understanding relative contributions of ecological and evolutionary variable on the focal processes. By using the chain rule, dynamics of an attribute (e.g., the per-capita growth rate of predator species), r, can be represented by

$$\frac{dr}{dt} = \underbrace{\frac{\partial r}{\partial n}\frac{dn}{dt}}_{\text{Eco}} + \underbrace{\frac{\partial r}{\partial z}\frac{dz}{dt}}_{\text{Evo}}, \tag{2.3}$$

where n and z are ecological (e.g., prey density) and evolutionary (e.g., prey defense trait) variables affecting r, respectively. See Appendix for examples of applications of the Geber method to theoretical models as like Eq. (2.2). The Geber method was further developed by Ellner et al. (2011) and Govaert et al. (2016) by incorporating the Price equation approach for understanding the effects of phenotypic plasticity as well as evolution of plasticity. See Govaert (2018) for the state of the art of the partitioning method.

Numerical simulations of Eq. (2.2) can indeed replicate the findings of previous studies: without evolution, the phase lag between predator and prey densities is quarter (Fig. 2.4a, b), but rapid prey evolution can cause the antiphase cycles

(Fig. 2.4c–e). When the cost of defense is small, predator–prey dynamics approaches to cryptic cycles (Fig. 2.4f–h). By using the Geber method, we can see that predator dynamics is strongly driven by the evolution of the prey defense trait in the antiphase cycles (Fig. 2.4d). When we take the absolute contributions of ecological and evolutionary variables in the long term, the evo/eco ratio is 2.94 in the antiphase cycles, indicating that the prey defense trait has stronger influence on predator dynamics than the prey density. On the other hand, the evo/eco ratio in cryptic cycles is 8.24 (Fig. 2.4g), which suggests that prey defense evolution is much more important than prey density dynamics for driving predator dynamics in cryptic cycles.

2.2.3 *Other Possible Cycles in Models*

Cortez and Weitz (2014) further demonstrated that coevolution between a predator offense trait and a prey defense trait can promote "clockwise" cycles, where the phase lag is three-quarters and predators increase first and then prey increase later (Figs. 2.3g, h and 2.4i–k). This results in clockwise cycles in a phase-plane plot (Fig. 2.4k) unlike the classical "counterclockwise" cycles (Fig. 2.4b). The pattern is very enigmatic as it looks as if the prey species exploit the predator species. The evo/eco ratio based on the Geber method is 1.74, suggesting that evolution is again more influential than ecology here. In addition to theoretical models (Fig. 2.4i–k), they found time-series data that looks like clockwise cycles in an experimental cyanophage-cyanobacterium system (Fig. 2.3g, h) (Cannon et al. 1976). However, no study has confirmed experimentally that coevolution can actually cause the clockwise cycles so far.

On the other hand, Yamamichi and Ellner (2016) showed that predator–prey coevolution can drive in-phase cycles, where prey and predator densities increase and decrease simultaneously (Fig. 2.5a–c). Therefore, everything seems possible in theory: the phase lag between predator and prey densities can be quarter (without evolution: Fig. 2.4b), half (with prey evolution: Fig. 2.4e), three-quarters (with coevolution: Fig. 2.4k), or zero (with coevolution: Fig. 2.5c). It will be interesting to manipulate genetic variation and examine whether rapid evolution can drive the clockwise and in-phase cycles in future studies.

2.2.4 *Does Rapid Evolution Stabilize or Destabilize Ecological Dynamics?*

Thus far, I have discussed the effects of rapid evolution on the characteristics of population cycles (e.g., the phase lag or cycle period), but another effect of rapid evolution is to stabilize (i.e., reduce the amplitude of limit cycles) or destabilize (i.e., increase the amplitude of limit cycles) predator–prey dynamics (Abrams 2000). For

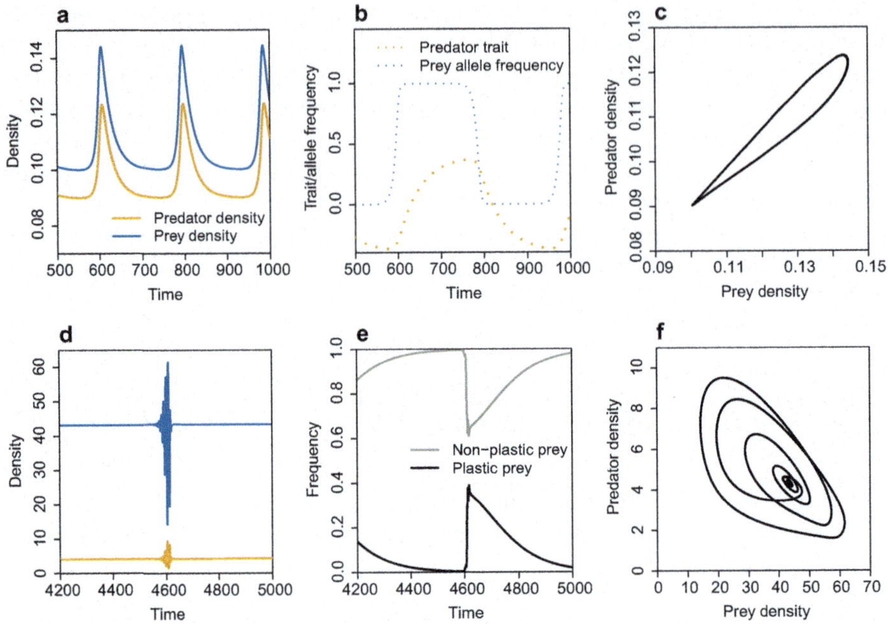

Fig. 2.5 In-phase cycles driven by coevolution (**a–c**) and intermittent cycles due to evolution of inducible defense (**d–f**). (**a**, **d**): Orange and blue lines represent predator and prey densities, respectively. (**b**): Dotted orange and blue lines represent mean predator trait and prey allele frequency, respectively. (**e**): Gray and black lines represent non-plastic and plastic prey frequencies, respectively. Models and parameters of **a–c** and **d–f** are based on Fig. 6b of Yamamichi and Ellner (2016) and Fig. 6b–d of Yamamichi et al. (2011), respectively

example, microcosm experiments of zooplankton and phytoplankton found that evolution can either stabilize (Steiner and Masse 2013) or destabilize (Becks et al. 2010) cycles.

From a theoretical perspective, previous studies have suggested that rapid evolution can cause a transition from a limit cycle to a stable equilibrium (or vice versa) through a Hopf bifurcation. This can be understood by considering a linearized system around the equilibrium using a Jacobian matrix based on Eq. (2.2):

$$\mathbf{J} = \begin{pmatrix} J_{11} & J_{12} & J_{13} \\ J_{21} & J_{22} & J_{23} \\ J_{31} & J_{32} & J_{33} \end{pmatrix} = \begin{pmatrix} x(f_x - g_x) & -xg_y & x(f_{\bar{\alpha}} - g_{\bar{\alpha}}) \\ yh_x & y(h_y - d_y) & yh_{\bar{\alpha}} \\ V(f_{\alpha x} - g_{\alpha x}) & -Vg_{\alpha y} & V(f_{\alpha\alpha} - g_{\alpha\alpha} + f_{\alpha\bar{\alpha}} - g_{\alpha\bar{\alpha}}) \end{pmatrix}, \tag{2.4}$$

where subscripts indicate partial differentiation (e.g., $f_x = \frac{\partial f}{\partial x}$). When there is no genetic variation, ecological subsystem (J_{11}, J_{12}, J_{21}, and J_{22}) determines the system stability. J_{33} is evolutionary subsystem, J_{13} and J_{23} represent the effects of evolution on ecological dynamics, and J_{31} and J_{32} show the effects of population densities on evolutionary dynamics. Cortez (2016) revealed that disruptive selection ($J_{33} > 0$) always destabilizes eco-evolutionary dynamics, whereas stabilizing evolution ($J_{33} < 0$) stabilizes or destabilizes dynamics. Therefore, it is important to understand the nature of selection pressure on traits to understand the (de)stabilizing effect of rapid evolution. Effects of predator evolution and coevolution on stability were further analyzed by Cortez and Patel (2017) and Cortez (2018), respectively, and it turned out that there is a counterintuitive outcome where destabilization occurs by the difference in the strength of the stabilities of the subsystems, i.e., subsets of the components of the whole system (Cortez 2018).

Patel et al. (2018) recently proposed a general approach for understanding effects of ecology and evolution on stability, and clarified that the evo-eco-evo feedback (i.e., effects of trait changes on selection via population dynamics) is critical for sufficiently slow evolutionary rates (when the additive genetic variance, V, is sufficiently small), whereas the eco-evo-eco feedback (i.e., effects of population densities on growth mediated by traits) is critical for sufficiently fast evolutionary rates (when the additive genetic variance, V, is sufficiently large). Based on the above approaches, Cortez et al. (2020) identified how ecological, evolutionary, and eco-evolutionary feedbacks influence stability of predator–prey models that are empirically parameterized. They found that evolutionary and eco-evolutionary feedbacks involving the interactions between prey species tended to be destabilizing, and eco-evolutionary feedbacks can alter system stability when evolution is fast or slow.

2.2.5 *Phenotypic Plasticity and Other Factors*

It is often difficult to distinguish rapid evolution and phenotypic plasticity, adaptive trait changes in response to environmental perturbations without altering genotype frequencies. For example, while green alga *Chlamydomonas* exhibits rapid evolution to be defended by clumping (Becks et al. 2010, 2012), another green alga, *Scenedesmus obliquus*, is known for clumping due to inducible defense, a kind of phenotypic plasticity: they sense chemicals released from predators (i.e., kairomone) and change their phenotypes to be defended (i.e., clumped) without changing genotype frequencies (Verschoor et al. 2004). In addition, phenotypic plasticity including adaptive behavior (Abrams 2010) can affect various ecological dynamics (Miner et al. 2005), and its importance has been recognized as a trait-mediated (indirect) interaction (TMI or TMII) in community ecology (Werner and Peacor 2003).

Microcosm experiments showed that inducible defense of *Scenedesmus* can stabilize predator–prey cycles (Verschoor et al. 2004), whereas theoretical studies have demonstrated that inducible defenses stabilize (Vos et al. 2004; Mougi and Kishida 2009) or destabilize (Kopp and Gabriel 2006; van Velzen et al. 2018) population dynamics. Several theoretical studies compared the effects of rapid evolution and inducible defenses on predator–prey dynamics and concluded that inducible defenses are better at stabilizing dynamics (Cortez 2011; Yamamichi et al. 2011, 2019b; Kovach-Orr and Fussmann 2013), because evolution responds to environmental changes with some time-delay and selection drives trait evolution along local fitness gradient, whereas inducible defense can cross fitness valleys and stabilize dynamics even under disruptive selection (Cortez 2011; Yamamichi et al. 2011, 2019b).

Furthermore, rapid evolution of inducible defense (i.e., changing frequencies of plastic and non-plastic genotypes) can cause intermittent cycles, where cycles are stabilized by the increase of the plastic genotype, and then destabilized by the increase of the non-plastic genotypes (as the plastic genotype has a fitness advantage in fluctuating environments: Fig. 2.7d–f). Here temporal fluctuations enable three prey types sharing a predator and a resource to coexist, and this may be an example of relative nonlinearity (Chesson 2000; Yamamichi and Letten 2020). Researchers examined the effects of evolution of inducible defense in rotifer–alga microcosm (Fischer et al. 2014), but the intermittent cycles were not observed so far. As there are several modeling approaches for inducible defenses, it is important to carefully choose a model for comparing theoretical predictions and empirical data (Yamamichi et al. 2019b).

Besides rapid evolution and inducible defenses, recent experimental studies examined the roles of trade-off shapes (Kasada et al. 2014), ecological stoichiometry (Declerck et al. 2015), coevolution (Hiltunen and Becks 2014; Frickel et al. 2016; Haafke et al. 2016), environmental fluctuations (Hiltunen et al. 2015), and abiotic stress (Hiltunen et al. 2018) in predator–prey systems as well as intraguild predation (Hiltunen et al. 2014a) and exploitative competition (Frickel et al. 2017) by using rotifer–alga, phage–alga, and ciliate–bacterium microcosm systems. In parallel, researchers are developing theory of eco-evolutionary dynamics (Govaert et al. 2019) by considering trade-off shapes (Ehrlich et al. 2017), ecological stoichiometry (Yamamichi et al. 2015), and coevolution (Mougi and Iwasa 2010, 2011; Mougi 2012; Tien and Ellner 2012; van Velzen and Gaedke 2017, 2018) in predator–prey systems as well as intraguild predation (Ellner and Becks 2011; Patel and Schreiber 2015), exploitative competition (van Velzen 2020), and apparent competition (Schreiber et al. 2011; Patel and Bürger 2019).

2.2.6 Evolutionary Rescue

When adaptive evolution occurs rapidly enough in the face of environmental deterioration, it may prevent extinction (Gomulkiewicz and Holt 1995): this is called evolutionary rescue (Kinnison and Hairston 2007; Alexander et al. 2014; Carlson et al. 2014; Bell 2017). This will be important for conservation and management of wild populations with rapid environmental changes due to human activities for understanding and predicting extinction dynamics. Evolutionary rescue has been studied in the context of medicine as well, because evolution of antibiotic-resistant bacteria is a major threat for human well-being (Alexander et al. 2014). Thus, while conservation biologists try to foster rescue, the goal in medicine and eradication of invasive species is preventing evolutionary rescue effectively. Previous studies employed theoretical and experimental approaches and demonstrated that evolutionary rescue is more likely to occur when the initial population size is large, the initial genetic variation is high, and the speed of environmental change is slow (Carlson et al. 2014).

While those studies tended to focus on single-species dynamics, an increasing number of studies found that interspecific interactions are important for understanding evolutionary rescue and species extinction in complex communities. For example, prey evolution may prevent predator extinction after the increase of the predator mortality (Fig. 2.6; Yamamichi and Miner 2015, Cortez and Yamamichi 2019). This "indirect evolutionary rescue" occurs because, when there is a trade-off between defense and growth in prey species (e.g., Yoshida et al. 2004; Meyer et al. 2006; Becks et al. 2010; Kasada et al. 2014), the increase of predator mortality drives adaptive evolution toward a less defended, rapidly growing prey phenotype. This results in the increased food quality for predator species and that can indirectly prevent predator extinction, as shown by the positive contribution of prey trait evolution to predator population growth (Fig. 2.6c). It is even possible for prey adaptation to increase the predator abundance after the increase of predator mortality (Fig. 2.6; Cortez and Yamamichi 2019). This counterintuitive phenomenon is called the hydra effect, as the hydra, the monster of Greek myth, increases the number of

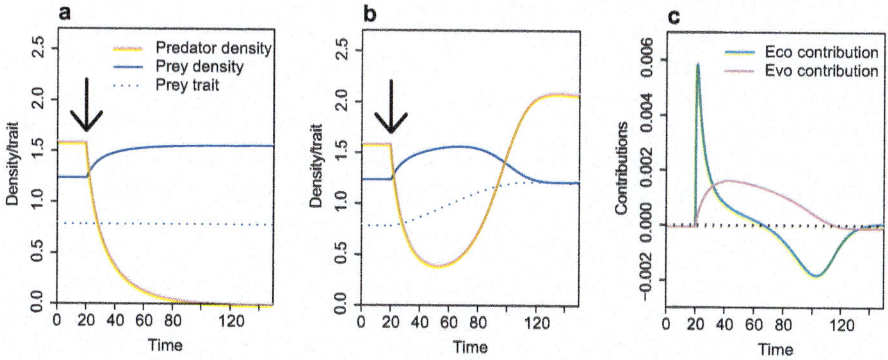

Fig. 2.6 Prey evolution can prevent predator extinction (indirect evolutionary rescue) and even increase predator density (hydra effect) when the predator mortality is increased (indicated by arrows). In **a** and **b**, orange and blue lines represent predator and prey densities, respectively. Dotted blue lines are the mean prey trait (i.e., the capture rate). In **c**, contributions of prey density and prey trait to predator per-capita growth rate are shown by green and magenta lines, respectively. See Appendix for details

necks when they are cut (reviewed in Abrams 2009, Schröder et al. 2014). The hydra effect occurs by various factors and makes it difficult to understand the consequences of increased mortality after environmental changes or harvesting.

Recent studies further clarified the effects of competition (Osmond and de Mazancourt 2013; Thompson and Fronhofer 2019), predation (Osmond et al. 2017), and mutualism (Jones and Gomulkiewicz 2012; Northfield and Ives 2013) on evolutionary rescue. Some experimental studies have examined "community rescue" by using soil microbial communities, showing that the community size and initial diversity are important for adapting to deteriorated environments (Low-Décarie et al. 2015). However, evolutionary rescue experiments with two or three species will be more informative for understanding which factors (e.g., interspecific interaction types and the presence of genetic variation) can change the outcome of eco-evolutionary extinction processes.

Evolutionary rescue will be critical for understanding stable coexistence of various species in community dynamics. If the growth rate of species when rare can be increased by adaptive evolution, species coexistence will be more likely and evolution will work as a stabilizing force (Lankau 2011) in the framework of modern coexistence theory (Chesson 2000). This can be seen as negative frequency-dependence in the community level due to rapid contemporary evolution. Species coexistence is enhanced with ecological character displacement of competing species (Germain et al. 2018) (but see Abrams 1987), with a trade-off between intraspecific and interspecific interactions in horizontal (i.e., competing) communities (Vasseur et al. 2011; Mougi 2013; Wittmann and Fukami 2018), with adaptive foraging of predators or adaptive defense by shared prey (Kondoh 2003; Schreiber et al. 2018; van Velzen 2020), or with density-dependent "intraspecific adaptation load," a reduction of population growth due to adaptation to intraspecific sexual or social interactions (Yamamichi et al. 2020). Recent experimental studies further revealed the role of adaptive evolution on promoting species coexistence (Hiltunen et al. 2017; Hart et al. 2019; Scheuerl et al. 2019).

In addition, adaptive evolution sometimes drives unintuitive dynamics where individual-level adaptation results in extinction: this is called evolutionary suicide (Parvinen 2005) or Darwinian extinction (Webb 2003). For example, selfish individuals can propagate and this may result in the tragedy of the commons, where persuading individual benefits results in the collapse of population benefits (Hardin 1968; Rankin et al. 2007). It is also possible for rapid evolution to promote extinction: in a coevolving predator–prey model, increased additive genetic variance (and the resultant faster evolution) of the predator trait can amplify predator–prey cycles and may result in extinction (Fig. 2.7; Yamamichi and Ellner 2016). Thus, adaptive evolution does not always work for preventing species extinction and coexistence, and its effects depend on various factors in eco-evolutionary dynamics.

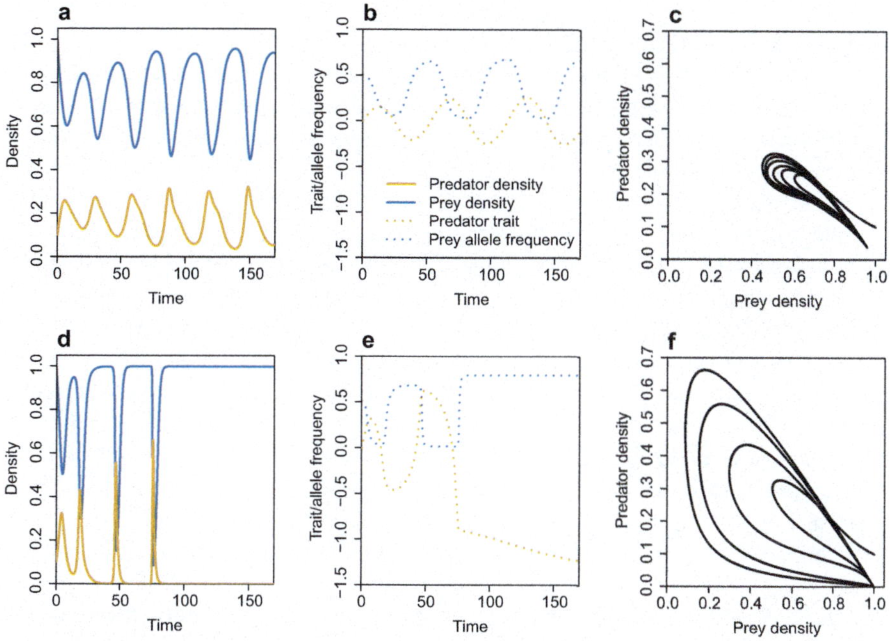

Fig. 2.7 Increasing the speed of evolution can promote extinction in coevolving predator–prey dynamics. In **a–c**, predator evolution is slow (additive genetic variance, V, is 0.01), whereas predator evolution is rapid ($V = 0.03$) and it results in extinction in **d–e**. In **a** and **d**, orange and blue lines represent predator and prey densities, respectively. In **b** and **e**, dotted orange and blue lines represent predator mean trait value and prey allele frequency, respectively. Models and parameters are based on Fig. 5 of Yamamichi and Ellner (2016)

Box 2.2: How to Quantify the Phase Lag Between Predator and Prey Densities

There are several approaches for investigating the phase lag between predator and prey densities. For example, some studies calculated the phase lag from stationary time-series (simulation) data by using the dominant frequency of the Fourier transform (Mougi 2012; van Velzen and Gaedke 2018). On the other hand, Hiltunen et al. (2014b) analyzed transient dynamics and employed an Evolutionary Dynamics Index (EDI) approach. When the transition from classic cycles to antiphase cycles occurs (Fig. 2.8a) by rapid evolution to be more defended (Fig. 2.8c, d), dynamics shifts from an oval limit cycle to a line with negative slope in a phase-plane plot (Fig. 2.8b). The transition in the phase-plane trajectories can be captured by inverse eccentricity, the ratio between the orthogonal long and short axes of an ellipse. EDI is defined as

(continued)

Box 2.2 (continued)
the difference between the first and last values of the fitted regression line of a sequence of inverse eccentricity values calculated for a moving window over the data series. Large negative EDI values indicate transition from quarter-period lag to antiphase cycles.

For quantifying the phase lag of ordinary differential equations, Jones and Ellner (2007) used the two prey genotype model (Eq. 2.1) and approximated eigenvectors at the bifurcation point. They found that the phase lag between the undefended prey genotype and predator is quarter, whereas that between the defended prey genotype and predator is three-quarter, and thus the total prey density and predator density shows antiphase cycles. On the other hand, several studies have examined the phase lag by focusing on the linearized system around the internal equilibrium using trigonometric functions (Bulmer 1975; Cortez and Ellner 2010; Cortez 2011). Bulmer (1975) assumed that the prey population oscillates about the equilibrium, $(\bar{x}, \bar{y})$, as

$$x - \bar{x} = A \sin(2\pi\omega t), \tag{2.5}$$

where A and ω are the amplitude and frequency of the oscillations. The linearized predator dynamics is then represented as

$$\frac{dy}{dt} = J_{21} A \sin(2\pi\omega t) + J_{22}(y - \bar{y}), \tag{2.6}$$

where J_{21} and J_{22} are components of the Jacobian matrix (Eq. 2.4). When J_{22} is zero as the Rosenzweig–MacArthur model (e.g., there is no interference between predators), $y = -B\cos(2\pi\omega t) = B\sin(2\pi\omega t - \pi/2)$ where $B = J_{21}A/(2\pi\omega)$. Thus, the phase lag is a quarter of a period ($\pi/2$) in the classic predator–prey cycles. This approach was further generalized in subsequent studies assuming a Hopf bifurcation (Ellner and Becks 2011; Cortez 2016, 2018; Cortez and Patel 2017).

2.3 Conclusions and Prospects

Eco-evolutionary dynamics is rapidly growing research area (Fig. 2.1, Table 2.1). In this chapter, I introduced experimental and theoretical studies on predator–prey cycles and evolutionary rescue. Those studies revealed that rapid prey evolution and coevolution can promote antiphase, cryptic, and clockwise cycles (Figs. 2.3 and 2.4). Furthermore, coevolution and evolution of phenotypic plasticity (i.e., frequency dynamics of genotypes with inducible defense) may drive in-phase and intermittent cycles, respectively (Fig. 2.5). Prey evolution can also prevent predator extinction (i.e., indirect evolutionary rescue) and increase the predator density in

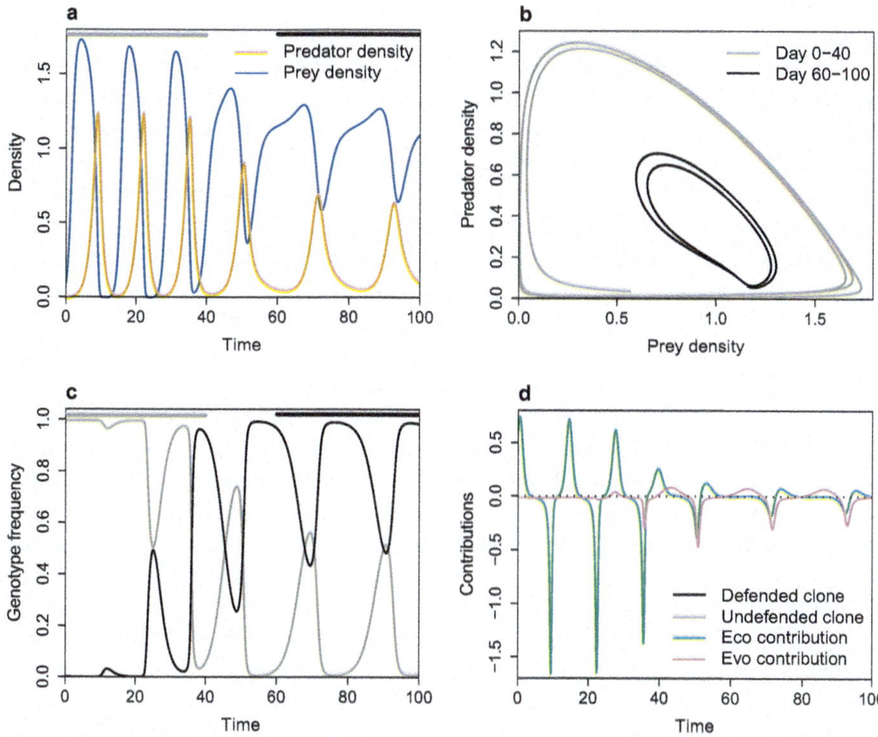

Fig. 2.8 Transient dynamics to antiphase cycles in the two clonal prey model. (**a**): Orange and blue lines represent predator and prey densities, respectively. (**b**): Gray and black lines are dynamics from $t = 0$ to 40 and from $t = 60$ to 100, respectively. (**c**): Gray and black lines represent dynamics of undefended and defended prey genotypes, respectively. (**d**): Green and magenta lines represent contributions of prey density (i.e., the ecological variable) and prey mean trait (i.e., the evolutionary variable) to predator per-capita growth rate, respectively. See Appendix for details

response to the increased predator mortality (i.e., hydra effect: Fig. 2.6). However, rapid evolution does not always prevent extinction: rapid predator evolution may result in predator extinction due to coevolutionary cycles (Fig. 2.7). While these theoretical and experimental studies have demonstrated the importance of rapid evolution for better understanding community dynamics, there is room for further investigation. Here I discuss future perspectives, focusing on (1) the mechanisms of adaptive phenotypic changes, (2) types of ecological interactions, (3) scales of communities, and (4) eco-evolutionary dynamics in the wild.

As I pointed out above, it is sometimes difficult to distinguish evolution and phenotypic plasticity. Even when adaptation is due to genetic evolution, there are many possible genetic architectures. For example, haploid and diploid genetic systems may have different consequences for eco-evolutionary dynamics (Schreiber et al. 2018). Coevolution between Mendelian (discrete) and quantitative (continuous) traits may drive distinct dynamics comparing to coevolution between

Mendelian traits or that between quantitative traits (Yamamichi and Ellner 2016) although there are parallel relationships between them (Yamamichi et al. 2019c). It is still unclear how epigenetics, epistasis, allele dominance, maternal effects, and recombination (Patel and Bürger 2019) affect eco-evolutionary dynamics. Several studies examined genomic aspects of eco-evolutionary dynamics (Becks et al. 2012; Hiltunen et al. 2018), and this kind of empirical data may be useful for predicting eco-evolutionary dynamics if there is a tight link between genetic architecture and dynamics (Matthews et al. 2018; Rudman et al. 2018).

While previous studies tended to focus on predator–prey systems, it will be interesting to investigate eco-evolutionary dynamics in competitive interactions (Lankau 2011; Vasseur et al. 2011; Mougi 2013; Wittmann and Fukami 2018; Yamamichi et al. 2020) as well as mutualistic interactions (Jones et al. 2009a; Northfield and Ives 2013). For example, Northfield and Ives (2013) showed that coevolution with antagonistic interactions (predation and competition) can buffer the effects of environmental changes, whereas that with mutualistic interactions amplify the environmental effects. The combination of various interactions may further promote complex dynamics (Chap. 8 of this book). In addition, adaptation to intraspecific interactions driven by sexual and social selection may cause interesting but largely neglected eco-evolutionary dynamics (Giery and Layman 2019; Svensson 2019; Yamamichi et al. 2020).

As Lawton (1999) suggested, it may be possible to tackle the effects of rapid evolution on community dynamics by avoiding intermediate biological/temporal/spatial scales, because even three species dynamics is very complex, as shown in systems with direct resource competition (Frickel et al. 2017) and with intraguild predation (Hiltunen et al. 2014a). For example, Kondoh (2003) showed that adaptive foraging makes the complexity–stability relationship positive and thus a large complex community can be stable (May 1972). Fronhofer and Altermatt (2015) showed how eco-evolutionary dynamics drive range expansion by using bacteria experiments. In addition to the above two studies with higher biological and spatial scales, trophic metacommunity dynamics with rapid evolution will be an interesting topic for future studies (Toju et al. 2017).

Finally, previous studies on eco-evolutionary dynamics tended to focus on simple laboratory systems and thus might overestimate the effects of evolution on dynamics. Hence, researchers have discussed the importance of studying eco-evolutionary dynamics in the wild (Hendry 2016, 2019; Toju et al. 2017). Recently Schaffner et al. (2019) demonstrated eco-evolutionary dynamics in Oneida lake, New York, by focusing on *Daphnia* adapting to seasonally changing phytoplankton species composition. While zooplankton is suitable for laboratory experiments, time-series analyses for detecting ongoing feedback may be useful for utilizing big data of population abundance and genomic data for larger animals (e.g., fish abundance estimated by environmental DNA: Chap. 9 of this book). However, it should be noted that in addition to deterministic selection pressures, evolutionary processes are affected by migration between local populations as well as genetic drift in small populations in nature (Lowe et al. 2017), and it may make eco-evolutionary dynamics further complicated and detecting eco-evolutionary feedbacks difficult. For understanding complex eco-evolutionary dynamics in nature, combining theoretical, statistical, and empirical approaches will be increasingly important and essential in future studies.

Acknowledgements I would like to thank A. Mougi, T. Yoshida, K. Lyberger, S. Patel, and P. A. Abrams for their helpful comments. M.Y. was supported by the Japan Society for the Promotion of Science (JSPS) Grant-in-Aid for Scientific Research (KAKENHI) 15H02642, 16K18618, 16H04846, 18H02509, and 19K16223.

Appendix

In Fig. 2.2 of Box 2.1, I used the Wright–Fisher model with haploid inheritance and generation overlap, where allele frequency dynamics is represented by a difference equation,

$$f(t+1) = (1-H)f(t) + \frac{Hw_1(t)f(t)}{w_1(t)f(t) + w_2(t)[1-f(t)]}, \tag{2.7}$$

where $f(t)$ and $1-f(t)$ are the frequency of allele A and a at time t, respectively, $w_1(t)$ and $w_2(t)$ are temporally fluctuating fitness of A and a at time t, respectively, and H is the hatching rate from dormant propagules ($0 < H \leq 1$). I assumed that w_1 and w_2 are log-normally distributed with mean (1.05, 1) and standard deviation, $\sqrt{0.15}$, on the log scale. As the geometric mean of w_1/w_2 is larger than one, the allele A fixes and the allele a goes extinct when $H = 1$ (i.e., no generation overlap).

In Figs. 2.4c–h and 2.6, I used a predator–prey model with prey evolution of Fig. 3a in Abrams and Matsuda (1997). Dynamics of the prey density, x, predator density, y, and the mean prey trait, $\overline{\alpha}$, are represented by

$$\begin{aligned}
\frac{dx}{dt} &= x\left(r_0 + r_1\alpha - \frac{x}{K} - \frac{\alpha y}{1+h\overline{\alpha}x}\right)\Big|_{\alpha=\overline{\alpha}}, \\
\frac{dy}{dt} &= y\left(\frac{c\overline{\alpha}x}{1+h\overline{\alpha}x} - d\right), \\
\frac{d\overline{\alpha}}{dt} &= V_x\left(r_1 - \frac{y}{1+h\overline{\alpha}x}\right),
\end{aligned} \tag{2.8}$$

where r_0 is the basic growth rate, r_1 is the coefficient of defense cost, K is carrying capacity, h is the handling time, c is the conversion efficiency, d is the predator mortality, and V_x is the additive genetic variance. Here the prey trait is under frequency-dependent selection. For the Geber method (Eq. 2.3), I assumed $r = dy/dt/y$, $n = x$, and $z = \overline{\alpha}$. Effects of the ecological and evolutionary variables on the predator per-capita growth rate are represented by

$$\begin{aligned}\frac{\partial r}{\partial n} &= \frac{c\bar{\alpha}}{(1+h\bar{\alpha}x)^2},\\ \frac{\partial r}{\partial z} &= \frac{cx}{(1+h\bar{\alpha}x)^2},\end{aligned} \tag{2.9}$$

Because an internal equilibrium is

$$\begin{aligned}\overline{x} &= \frac{d}{(c-hd)\overline{\alpha}},\\ \overline{y} &= \frac{(r_0K + r_1\overline{\alpha}K - \overline{x})(1+h\overline{x}\overline{\alpha})}{\overline{\alpha}K},\end{aligned} \tag{2.10}$$

and, when the trait evolves, this can be simplified as

$$\begin{aligned}\widehat{x} &= r_0K,\\ \widehat{y} &= \frac{cr_1}{c-hd},\\ \widehat{\overline{\alpha}} &= \frac{d}{r_0K(c-hd)},\end{aligned} \tag{2.11}$$

the predator density can increase with the increased mortality ($\frac{\partial\widehat{y}}{\partial d} > 0$) with evolution (i.e., the hydra effect) while it is a decreasing function of mortality ($\frac{\partial\overline{y}}{\partial d} < 0$) without evolution.

The Jacobian matrix is

$$\mathbf{J} = \begin{pmatrix} \dfrac{r_1hd^2/K + chdr_0^2 - c^2r_0^2}{cr_0(c-hd)} & -\dfrac{d}{c} & \dfrac{r_0r_1hKd}{c} \\ \dfrac{r_1d}{r_0K} & 0 & r_0r_1K(c-hd) \\ V_x\dfrac{hr_1d}{cr_0K} & -V_x\dfrac{c-hd}{c} & V_x\dfrac{r_0r_1hK(c-hd)}{c} \end{pmatrix}, \tag{2.12}$$

where I assume $c - hd > 0$. The sign structure of the Jacobian matrix is

$$\begin{pmatrix} \pm & - & + \\ + & 0 & + \\ + & - & + \end{pmatrix}, \tag{2.13}$$

and thus there is destabilizing selection for the prey trait ($J_{33} > 0$).

In Fig. 2.4c–e, $r_0 = 2.5$, $r_1 = 0.8$, $K = h = c = 1$, $d = 0.5$, and $V_x = 0.05$. In Fig. 2.4f–h, $r_1 = 0.1$ and other parameter values are the same as Fig. 2.4c–e. In Fig. 2.6, $K = 0.5$ and d is changed from 0.5 to 0.6. The additive genetic variance is zero in Fig. 2.6a and $V_x = 0.01$ in Fig. 2.6b, c. In Fig. 2.4a, b, I assumed that the prey trait does not evolve, and parameter values are $r_0 = 1.7$, $r_1 = V_x = 0$, $K = c = d = 1$, $h = 0.5$, and $\overline{\alpha}(0) = 3.6$.

In Fig. 2.4i–k, I used a predator–prey model with coevolution of Fig. 5b in Cortez and Weitz (2014). Dynamics of the prey density, x, the predator density, y, the mean prey trait, $\overline{\alpha}$, and the mean prey trait, $\overline{\beta}$, are represented by

$$
\begin{aligned}
\frac{dx}{dt} &= x\left[(r_0 + r_1\alpha)\left(1 - \frac{x}{k_0 + k_1\overline{\alpha}}\right) - \left(g_0 + g_1\alpha + g_2\overline{\beta} + g_3\alpha\overline{\beta}\right)\frac{y}{1 + hx}\right]\Bigg|_{\alpha=\overline{\alpha}},\\
\frac{dy}{dt} &= y\left[\left(g_0 + g_1\overline{\alpha} + g_2\beta + g_3\overline{\alpha}\beta\right)\frac{cx}{1 + hx} - (d_0 + d_1\beta)y^{\frac{1}{2}}\right]\Bigg|_{\beta=\overline{\beta}},\\
\frac{d\overline{\alpha}}{dt} &= V_x\overline{\alpha}(1 - \overline{\alpha})\left[r_1\left(1 - \frac{x}{k_0 + k_1\overline{\alpha}}\right) - \left(g_1 + g_3\overline{\beta}\right)\frac{y}{1 + hx}\right],\\
\frac{d\overline{\beta}}{dt} &= V_y\overline{\beta}\left(1 - \overline{\beta}\right)\left[\left(g_2 + g_3\overline{\alpha}\right)\frac{cx}{1 + hx} - d_1 y^{\frac{1}{2}}\right],
\end{aligned}
\tag{2.14}
$$

where k_0 is the basic carrying capacity, k_1 is the coefficient of defense cost for carrying capacity, g_0 is the basic attack rate, g_1, g_2, and g_3 are the coefficient of traits on the attack rate, d_0 is the basic mortality, d_1 is the coefficient of offense cost for mortality, and V_y is the additive genetic variance of the predator offense trait. Here selection of the prey trait is frequency-dependent, whereas selection of the predator trait is frequency-independent. For the Geber method (Eq. 2.3), I assumed $X = dy/dt/y$, $k = x$, and $z = \overline{\alpha}$. Effects of the ecological and evolutionary variables on the predator per-capita growth rate are represented by

$$
\begin{aligned}
\frac{\partial r}{\partial n} &= \frac{c\left(g_0 + g_1\overline{\alpha} + g_2\overline{\beta} + g_3\overline{\alpha}\overline{\beta}\right)}{(1 + hx)^2},\\
\frac{\partial r}{\partial z} &= \frac{cx\left(g_1 + g_3\overline{\beta}\right)}{1 + hx},
\end{aligned}
\tag{2.15}
$$

where I assumed $r_0 = 2.5$, $r_1 = 3.5$, $k_0 = 1$, $k_1 = 0.1$, $g_0 = 0.65$, $g_1 = 3$, $g_2 = 2.3$, $g_3 = -0.2$, $c = 1.7$, $d_0 = 0.76$, $d_1 = 1.77$, and $h = V_x = V_y = 1$.

In Fig. 2.8 of Box 2.2, I used a model with one predator and two prey genotypes with prey logistic growth and the Holling type II functional response (Abrams and Matsuda 1997; Yoshida et al. 2007; Yamamichi et al. 2014). Dynamics of the undefended prey density, x_1, the defended prey density, x_2, and the predator density, y, are represented by

$$\frac{dx_i}{dt} = x_i\left(r_i - X - \frac{p_i g y}{k+Q}\right), \quad i = 1, 2,$$
$$\frac{dy}{dt} = y\left(\frac{gQ}{k+Q} - d\right), \tag{2.16}$$

where $X = x_1 + x_2$, $Q = p_1x_1 + p_2x_2$, r_i is the intrinsic growth rate of prey genotype i, p_i is the food value of prey genotype i with $0 \le p_i \le 1$, g is the maximum attack rate, k is the half saturation constant of predation, and d is the predator mortality. I assumed $r_1 = 1.8, r_2 = 1.2, p_1 = 1, p_2 = 0.18, g = 2, k = 0.3$, and $d = 1$. The initial condition is $(x_1, x_2, y) = (0.1, 0.0001, 0.01)$. As for the Geber method, I considered the effects of total prey density (X) and mean food value ($\overline{p} = Q/X$) on the predator per-capita growth rate, $dy/dt/y$. Each component of the Geber method is represented by

$$\frac{\partial r}{\partial n} = \frac{gk\overline{p}}{(k+\overline{p}X)^2},$$
$$\frac{dn}{dt} = \frac{dx_1}{dt} + \frac{dx_2}{dt},$$
$$\frac{\partial r}{\partial z} = \frac{gkX}{(k+\overline{p}X)^2}, \tag{2.17}$$
$$\frac{dz}{dt} = \frac{p_1 - p_2}{X^2}\left(x_2\frac{dx_1}{dt} + x_1\frac{dx_2}{dt}\right).$$

References

Abrams PA (1987) Alternative models of character displacement and niche shift. I. Adaptive shifts in resource use when there is competition for nutritionally nonsubstitutable resources. Evolution 41:651–661

Abrams PA (1999) Is predator-mediated coexistence possible in unstable systems? Ecology 80:608–621

Abrams PA (2000) The evolution of predator-prey interactions: theory and evidence. Annu Rev Ecol Syst 31:79–105

Abrams PA (2001) Modelling the adaptive dynamics of traits involved in inter- and intraspecific interactions: an assessment of three methods. Ecol Lett 4:166–175

Abrams PA (2009) When does greater mortality increase population size? The long history and diverse mechanisms underlying the hydra effect. Ecol Lett 12:462–474

Abrams PA (2010) Implications of flexible foraging for interspecific interactions: lessons from simple models. Funct Ecol 24:7–17

Abrams PA, Matsuda H (1997) Prey adaptation as a cause of predator-prey cycles. Evolution 51:1742–1750

Alexander HK, Martin G, Martin OY, Bonhoeffer S (2014) Evolutionary rescue: linking theory for conservation and medicine. Evol Appl 7:1161–1179

Antonovics J (1992) Toward community genetics. In: Fritz RS, Simms EL (eds) Plant resistance to herbivores and pathogens: ecology, evolution, and genetics. University of Chicago Press, Chicago, IL, pp 426–449

Becks L, Ellner SP, Jones LE, Hairston NG Jr (2010) Reduction of adaptive genetic diversity radically alters eco-evolutionary community dynamics. Ecol Lett 13:989–997

Becks L, Ellner SP, Jones LE, Hairston NG Jr (2012) The functional genomics of an eco-evolutionary feedback loop: linking gene expression, trait evolution, and community dynamics. Ecol Lett 15:492–501

Bell G (2010) Fluctuating selection: the perpetual renewal of adaptation in variable environments. Philos Trans R Soc B Biol Sci 365:87–97

Bell G (2017) Evolutionary rescue. Annu Rev Ecol Evol Syst 48:605–627

Bergland AO, Behrman EL, O'Brien KR, Schmidt PS, Petrov DA (2014) Genomic evidence of rapid and stable adaptive oscillations over seasonal time scales in *Drosophila*. PLoS Genet 10: e1004775

Bertram J, Masel J (2019) Different mechanisms drive the maintenance of polymorphism at loci subject to strong versus weak fluctuating selection. Evolution 73:883–896

Bohannan BJM, Lenski RE (1999) Effect of prey heterogeneity on the response of a model food chain to resource enrichment. Am Nat 153:73–82

Bolnick DI, Amarasekare P, Araújo MS, Bürger R, Levine JM, Novak M, Rudolf VHW, Schreiber SJ, Urban MC, Vasseur DA (2011) Why intraspecific trait variation matters in community ecology. Trends Ecol Evol 26:183–192

Bulmer MG (1975) Phase relations in the ten-year cycle. J Anim Ecol 44:609–621

Cannon RE, Shane MS, Whitaker JM (1976) Interaction of *Plectonema boryanum* (Cyanophyceae) and the LPP-cyanophages in continuous culture. J Phycol 12:418–421

Carlson SM, Cunningham CJ, Westley PAH (2014) Evolutionary rescue in a changing world. Trends Ecol Evol 29:521–530

Chesson P (2000) Mechanisms of maintenance of species diversity. Annu Rev Ecol Syst 31:343–366

Chitty D (1960) Population processes in the vole and their relevance to general theory. Can J Zool 38:99–113

Cortez MH (2011) Comparing the qualitatively different effects rapidly evolving and rapidly induced defences have on predator-prey interactions. Ecol Lett 14:202–209

Cortez MH (2016) How the magnitude of prey genetic variation alters predator-prey eco-evolutionary dynamics. Am Nat 188:329–341

Cortez MH (2018) Genetic variation determines which feedbacks drive and alter predator-prey eco-evolutionary cycles. Ecol Monogr 88:353–371

Cortez MH, Ellner SP (2010) Understanding rapid evolution in predator-prey interactions using the theory of fast-slow dynamical systems. Am Nat 176:E109–E127

Cortez MH, Patel S (2017) The effects of predator evolution and genetic variation on predator–prey population-level dynamics. Bull Math Biol 79:1510–1538

Cortez MH, Weitz JS (2014) Coevolution can reverse predator–prey cycles. Proc Natl Acad Sci U S A 111:7486–7491

Cortez MH, Yamamichi M (2019) How (co)evolution alters predator responses to increased mortality: extinction thresholds and hydra effects. Ecology 100:e02789

Cortez MH, Patel S, Schreiber SJ (2020) Destabilizing evolutionary and eco-evolutionary feedbacks drive empirical eco-evolutionary cycles. Proc R Soc B Biol Sci 287:20192298

Darwin C (1859) On the origin of species by means of natural selection, or the preservation of favoured races in the struggle for life. John Murray, London

de Roos AM, Persson L (2013) Population and community ecology of ontogenetic development. Princeton University Press, Princeton, NJ

Declerck SAJ, Malo AR, Diehl S, Waasdorp D, Lemmen KD, Proios K, Papakostas S (2015) Rapid adaptation of herbivore consumers to nutrient limitation: eco-evolutionary feedbacks to population demography and resource control. Ecol Lett 18:553–562

Dempster ER (1955) Maintenance of genetic heterogeneity. Cold Spring Harbor Symposia on Quantitative Biology 20:25–32

Des Roches S, Post DM, Turley NE, Bailey JK, Hendry AP, Kinnison MT, Schweitzer JA, Palkovacs EP (2018) The ecological importance of intraspecific variation. Nat Ecol Evol 2:57–64

Ehrlich E, Becks L, Gaedke U (2017) Trait-fitness relationships determine how trade-off shapes affect species coexistence. Ecology 98:3188–3198

Ellner SP (2013) Rapid evolution: from genes to communities, and back again? Funct Ecol 27:1087–1099

Ellner SP, Becks L (2011) Rapid prey evolution and the dynamics of two-predator food webs. Theor Ecol 4:133–152

Ellner S, Hairston NG Jr (1994) Role of overlapping generations in maintaining genetic variation in a fluctuating environment. Am Nat 143:403–417

Ellner SP, Geber MA, Hairston NG Jr (2011) Does rapid evolution matter? Measuring the rate of contemporary evolution and its impacts on ecological dynamics. Ecol Lett 14:603–614

Fischer BB, Kwiatkowski M, Ackermann M, Krismer J, Roffler S, Suter MJF, Eggen RIL, Matthews B (2014) Phenotypic plasticity influences the eco-evolutionary dynamics of a predator–prey system. Ecology 95:3080–3092

Frickel J, Sieber M, Becks L (2016) Eco-evolutionary dynamics in a coevolving host–virus system. Ecol Lett 19:450–459

Frickel J, Theodosiou L, Becks L (2017) Rapid evolution of hosts begets species diversity at the cost of intraspecific diversity. Proc Natl Acad Sci U S A 114:11193–11198

Fronhofer EA, Altermatt F (2015) Eco-evolutionary feedbacks during experimental range expansions. Nat Commun 6:6844

Fussmann GF, Ellner SP, Shertzer KW, Hairston NG Jr (2000) Crossing the Hopf bifurcation in a live predator-prey system. Science 290:1358–1360

Fussmann GF, Ellner SP, Hairston NG Jr, Jones LE, Shertzer KW, Yoshida T (2005) Ecological and evolutionary dynamics of experimental plankton communities. Adv Ecol Res 37:221–243

Fussmann GF, Loreau M, Abrams PA (2007) Eco-evolutionary dynamics of communities and ecosystems. Funct Ecol 21:465–477

Germain RM, Williams JL, Schluter D, Angert AL (2018) Moving character displacement beyond characters using contemporary coexistence theory. Trends Ecol Evol 33:74–84

Giery ST, Layman CA (2019) Ecological consequences of sexually selected traits: an eco-evolutionary perspective. Q Rev Biol 94:29–74

Gingerich PD (1993) Quantification and comparison of evolutionary rates. Am J Sci 293A:453–478

Gomulkiewicz R, Holt RD (1995) When does evolution by natural selection prevent extinction? Evolution 49:201–207

Govaert L (2018) Eco-evolutionary partitioning metrics: a practical guide for biologists. Belg J Zool 148:167–202

Govaert L, Pantel JH, De Meester L (2016) Eco-evolutionary partitioning metrics: assessing the importance of ecological and evolutionary contributions to population and community change. Ecol Lett 19:839–853

Govaert L, Fronhofer EA, Lion S, Eizaguirre C, Bonte D, Egas M, Hendry AP, Martins ADB, Melián CJ, Raeymaekers JAM, Ratikainen II, Saether BE, Schweitzer JA, Matthews B (2019) Eco-evolutionary feedbacks—theoretical models and perspectives. Funct Ecol 33:13–30

Gulisija D, Kim Y, Plotkin JB (2016) Phenotypic plasticity promotes balanced polymorphism in periodic environments by a genomic storage effect. Genetics 202:1437–1448

Haafke J, Abou Chakra M, Becks L (2016) Eco-evolutionary feedback promotes Red Queen dynamics and selects for sex in predator populations. Evolution 70:641–652

Hairston NG Jr, Ellner SP, Geber MA, Yoshida T, Fox JA (2005) Rapid evolution and the convergence of ecological and evolutionary time. Ecol Lett 8:1114–1127

Haldane JBS, Jayakar SD (1963) Polymorphism due to selection of varying direction. J Genet 58:237–242

Hardin G (1968) The tragedy of the commons. Science 162:1243–1248
Hart SP, Turcotte MM, Levine JM (2019) Effects of rapid evolution on species coexistence. Proc Natl Acad Sci U S A 116:2112–2117
Hedrick PW (2006) Genetic polymorphism in heterogeneous environments: the age of genomics. Annu Rev Ecol Evol Syst 37:67–93
Hendry AP (2016) Eco-evolutionary dynamics. Princeton University Press, Princeton, NJ
Hendry AP (2019) A critique for eco-evolutionary dynamics. Funct Ecol 33:84–94
Hendry AP, Kinnison MT (1999) The pace of modern life: measuring rates of contemporary microevolution. Evolution 53:1637–1653
Hiltunen T, Becks L (2014) Consumer co-evolution as an important component of the eco-evolutionary feedback. Nat Commun 5:5226
Hiltunen T, Ellner SP, Hooker G, Jones LE, Hairston NG Jr (2014a) Eco-evolutionary dynamics in a three-species food web with intraguild predation: intriguingly complex. Adv Ecol Res 50:41–73
Hiltunen T, Hairston NG Jr, Hooker G, Jones LE, Ellner SP (2014b) A newly discovered role of evolution in previously published consumer–resource dynamics. Ecol Lett 17:915–923
Hiltunen T, Ayan GB, Becks L (2015) Environmental fluctuations restrict eco-evolutionary dynamics in predator–prey system. Proc R Soc B Biol Sci 282:20150013
Hiltunen T, Kaitala V, Laakso J, Becks L (2017) Evolutionary contribution to coexistence of competitors in microbial food webs. Proc R Soc B Biol Sci 284:20170415
Hiltunen T, Cairns J, Frickel J, Jalasvuori M, Laakso J, Kaitala V, Künzel S, Karakoc E, Becks L (2018) Dual-stressor selection alters eco-evolutionary dynamics in experimental communities. Nat Ecol Evol 2:1974–1981
Hoekstra HE, Hoekstra JM, Berrigan D, Vignieri SN, Hoang A, Hill CE, Beerli P, Kingsolver JG (2001) Strength and tempo of directional selection in the wild. Proc Natl Acad Sci U S A 98:9157–9160
Holt RD, Grover J, Tilman D (1994) Simple rules for interspecific dominance in systems with exploitative and apparent competition. Am Nat 144:741–771
Johnson MTJ, Stinchcombe JR (2007) An emerging synthesis between community ecology and evolutionary biology. Trends Ecol Evol 22:250–257
Jones LE, Ellner SP (2004) Evolutionary tradeoff and equilibrium in an aquatic predator-prey system. Bull Math Biol 66:1547–1573
Jones LE, Ellner SP (2007) Effects of rapid prey evolution on predator-prey cycles. J Math Biol 55:541–573
Jones EI, Gomulkiewicz R (2012) Biotic interactions, rapid evolution, and the establishment of introduced species. Am Nat 179:E28–E36
Jones EI, Ferrière R, Bronstein JL (2009a) Eco-evolutionary dynamics of mutualists and exploiters. Am Nat 174:780–794
Jones LE, Becks L, Ellner SP, Hairston NG Jr, Yoshida T, Fussmann GF (2009b) Rapid contemporary evolution and clonal food web dynamics. Philos Trans R Soc B Biol Sci 364:1579–1591
Kasada M, Yamamichi M, Yoshida T (2014) Form of an evolutionary tradeoff affects eco-evolutionary dynamics in a predator-prey system. Proc Natl Acad Sci U S A 111:16035–16040
Kendall BE, Prendergast J, Bjørnstad ON (1998) The macroecology of population dynamics: taxonomic and biogeographic patterns in population cycles. Ecol Lett 1:160–164
Kingsolver JG, Hoekstra HE, Hoekstra JM, Berrigan D, Vignieri SN, Hill CE, Hoang A, Gibert P, Beerli P (2001) The strength of phenotypic selection in natural populations. Am Nat 157:245–261
Kinnison MT, Hairston NG Jr (2007) Eco-evolutionary conservation biology: contemporary evolution and the dynamics of persistence. Funct Ecol 21:444–454
Koch H, Frickel J, Valiadi M, Becks L (2014) Why rapid, adaptive evolution matters for community dynamics. Front Ecol Evol 2:17

Kondoh M (2003) Foraging adaptation and the relationship between food-web complexity and stability. Science 299:1388–1391
Kopp M, Gabriel W (2006) The dynamic effects of an inducible defense in the Nicholson-Bailey model. Theor Popul Biol 70:43–55
Kovach-Orr C, Fussmann GF (2013) Evolutionary and plastic rescue in multitrophic model communities. Philos Trans R Soc B Biol Sci 368:20120084
Kretzschmar M, Nisbet RM, McCauley E (1993) A predator-prey model for zooplankton grazing on competing algal populations. Theor Popul Biol 44:32–66
Lankau RA (2011) Rapid evolutionary change and the coexistence of species. Annu Rev Ecol Evol Syst 42:335–354
Lawton JH (1999) Are there general laws in ecology? Oikos 84:177–192
Lotka AJ (1925) Elements of physical biology. Williams & Wilkins Company, Baltimore, MD
Low-Décarie E, Kolber M, Homme P, Lofano A, Dumbrell A, Gonzalez A, Bell G (2015) Community rescue in experimental metacommunities. Proc Natl Acad Sci U S A 112:14307–14312
Lowe WH, Kovach RP, Allendorf FW (2017) Population genetics and demography unite ecology and evolution. Trends Ecol Evol 32:141–152
Matthews B, Narwani A, Hausch S, Nonaka E, Peter H, Yamamichi M, Sullam KE, Bird KC, Thomas MK, Hanley TC, Turner CB (2011) Toward an integration of evolutionary biology and ecosystem science. Ecol Lett 14:690–701
Matthews B, Best RJ, Feulner PGD, Narwani A, Limberger R (2018) Evolution as an ecosystem process: insights from genomics. Genome 61:298–309
May RM (1972) Will a large complex system be stable? Nature 238:413–414
Messer PW, Ellner SP, Hairston NG Jr (2016) Can population genetics adapt to rapid evolution? Trends Genet 32:408–418
Meyer JR, Ellner SP, Hairston NG Jr, Jones LE, Yoshida T (2006) Prey evolution on the time scale of predator-prey dynamics revealed by allele-specific quantitative PCR. Proc Natl Acad Sci U S A 103:10690–10695
Miner BG, Sultan SE, Morgan SG, Padilla DK, Relyea RA (2005) Ecological consequences of phenotypic plasticity. Trends Ecol Evol 20:685–692
Mougi A (2012) Predator-prey coevolution driven by size selective predation can cause anti-synchronized and cryptic population dynamics. Theor Popul Biol 81:113–118
Mougi A (2013) Allelopathic adaptation can cause competitive coexistence. Theor Ecol 6:165–171
Mougi A, Iwasa Y (2010) Evolution towards oscillation or stability in a predator-prey system. Proc R Soc B Biol Sci 277:3163–3171
Mougi A, Iwasa Y (2011) Unique coevolutionary dynamics in a predator–prey system. J Theor Biol 277:83–89
Mougi A, Kishida O (2009) Reciprocal phenotypic plasticity can lead to stable predator-prey interaction. J Anim Ecol 78:1172–1181
Northfield TD, Ives AR (2013) Coevolution and the effects of climate change on interacting species. PLoS Biol 11:e1001685
Osmond MM, de Mazancourt C (2013) How competition affects evolutionary rescue. Philos Trans R Soc B Biol Sci 368:20120085
Osmond MM, Otto SP, Klausmeier CA (2017) When predators help prey adapt and persist in a changing environment. Am Nat 190:83–98
Parvinen K (2005) Evolutionary suicide. Acta Biotheor 53:241–264
Patel S, Bürger R (2019) Eco-evolutionary feedbacks between prey densities and linkage disequilibrium in the predator maintain diversity. Evolution 73:1533–1548
Patel S, Schreiber SJ (2015) Evolutionarily driven shifts in communities with intraguild predation. Am Nat 186:E98–E110
Patel S, Cortez MH, Schreiber SJ (2018) Partitioning the effects of eco-evolutionary feedbacks on community stability. Am Nat 191:381–394

Pimentel D (1961) Animal population regulation by the genetic feed-back mechanism. Am Nat 95:65–79

Pimentel D (1968) Population regulation and genetic feedback. Science 159:1432–1437

R Core Team (2019) R: a language and environment for statistical computing. R Foundation for Statistical Computing, Vienna, Austria

Rankin DJ, Bargum K, Kokko H (2007) The tragedy of the commons in evolutionary biology. Trends Ecol Evol 22:643–651

Reinhold K (2000) Maintenance of a genetic polymorphism by fluctuating selection on sex-limited traits. J Evol Biol 13:1009–1014

Reznick DN (2013) A critical look at reciprocity in ecology and evolution: introduction to the symposium. Am Nat 181:S1–S8

Rosenzweig ML, MacArthur RH (1963) Graphical representation and stability conditions of predator-prey interactions. Am Nat 97:209–223

Rudman SM, Barbour MA, Csilléry K, Gienapp P, Guillaume F, Hairston NG Jr, Hendry AP, Lasky JR, Rafajlović M, Räsänen K, Schmidt PS, Seehausen O, Therkildsen NO, Turcotte MM, Levine JM (2018) What genomic data can reveal about eco-evolutionary dynamics. Nat Ecol Evol 2:9–15

Schaffner LR, Govaert L, De Meester L, Ellner SP, Fairchild E, Miner BE, Rudstam LG, Spaak P, Hairston NG Jr (2019) Consumer-resource dynamics is an eco-evolutionary process in a natural plankton community. Nat Ecol Evol 3:1351–1358

Scheuerl T, Cairns J, Becks L, Hiltunen T (2019) Predator coevolution and prey trait variability determine species coexistence. Proc R Soc B Biol Sci 286:20190245

Schoener TW (2011) The newest synthesis: understanding the interplay of evolutionary and ecological dynamics. Science 331:426–429

Schreiber SJ, Bürger R, Bolnick DI (2011) The community effects of phenotypic and genetic variation within a predator population. Ecology 92:1582–1593

Schreiber SJ, Patel S, terHorst C (2018) Evolution as a coexistence mechanism: does genetic architecture matter? Am Nat 191:407–420

Schröder A, van Leeuwen A, Cameron TC (2014) When less is more: positive population-level effects of mortality. Trends Ecol Evol 29:614–624

Shefferson RP, Salguero-Gómez R (2015) Eco-evolutionary dynamics in plants: interactive processes at overlapping time-scales and their implications. J Ecol 103:789–797

Shertzer KW, Ellner SP, Fussmann GF, Hairston NG Jr (2002) Predator-prey cycles in an aquatic microcosm: testing hypotheses of mechanism. J Anim Ecol 71:802–815

Slobodkin LB (1961) Growth and regulation of animal populations. Holt, Rinehart and Winston, New York, NY

Steiner CF, Masse J (2013) The stabilizing effects of genetic diversity on predator-prey dynamics. F1000 Res 2:43

Svensson EI (2019) Eco-evolutionary dynamics of sexual selection and sexual conflict. Funct Ecol 33:60–72

Thompson JN (1998) Rapid evolution as an ecological process. Trends Ecol Evol 13:329–332

Thompson PL, Fronhofer EA (2019) The conflict between adaptation and dispersal for maintaining biodiversity in changing environments. Proc Natl Acad Sci U S A 116:21061–21067

Tien RJ, Ellner SP (2012) Variable cost of prey defense and coevolution in predator-prey systems. Ecol Monogr 82:491–504

Tilman D (1982) Resource competition and community structure. Princeton University Press, Princeton, NJ

Toju H, Yamamichi M, Guimarães PR Jr, Olesen JM, Mougi A, Yoshida T, Thompson JN (2017) Species-rich networks and eco-evolutionary synthesis at the metacommunity level. Nat Ecol Evol 1:0024

van Velzen E (2020) Predator coexistence through emergent fitness equalization. Ecology 101: e02995 van Velzen E, Gaedke U (2017) Disentangling eco-evolutionary dynamics of predator-prey coevolution: the case of antiphase cycles. Sci Rep 7:17125

van Velzen E, Gaedke U (2018) Reversed predator–prey cycles are driven by the amplitude of prey oscillations. Ecol Evol 8:6317–6329

van Velzen E, Thieser T, Berendonk T, Weitere M, Gaedke U (2018) Inducible defense destabilizes predator–prey dynamics: the importance of multiple predators. Oikos 127:1551–1562

Vasseur DA, Amarasekare P, Rudolf VHW, Levine JM (2011) Eco-evolutionary dynamics enable coexistence via neighbor-dependent selection. Am Nat 178:E96–E109

Verschoor AM, Vos M, van der Stap I (2004) Inducible defences prevent strong population fluctuations in bi- and tritrophic food chains. Ecol Lett 7:1143–1148

Volterra V (1926) Variazioni e fluttuazioni del numero d'individui in specie animali conviventi. Memorie della R Accademia Nazionale dei Lincei 6:30–113

Vos M, Kooi BW, DeAngelis DL, Mooij WM (2004) Inducible defences and the paradox of enrichment. Oikos 105:471–480

Webb C (2003) A complete classification of Darwinian extinction in ecological interactions. Am Nat 161:181–205

Werner EE, Peacor SD (2003) A review of trait-mediated indirect interactions in ecological communities. Ecology 84:1083–1100

Whitham TG, Bailey JK, Schweitzer JA, Shuster SM, Bangert RK, LeRoy CJ, Lonsdorf EV, Allan GJ, DiFazio SP, Potts BM, Fischer DG, Gehring CA, Lindroth RL, Marks JC, Hart SC, Wimp GM, Wooley SC (2006) A framework for community and ecosystem genetics: from genes to ecosystems. Nat Rev Genet 7:510–523

Wittmann MJ, Fukami T (2018) Eco-evolutionary buffering: rapid evolution facilitates regional species coexistence despite local priority effects. Am Nat 191:E171–E184

Wittmann MJ, Bergland AO, Feldman MW, Schmidt PS, Petrov DA (2017) Seasonally fluctuating selection can maintain polymorphism at many loci via segregation lift. Proc Natl Acad Sci U S A 114:E9932–E9941

Yamamichi M, Ellner SP (2016) Antagonistic coevolution between quantitative and Mendelian traits. Proc R Soc B Biol Sci 283:20152926

Yamamichi M, Hoso M (2017) Roles of maternal effects in maintaining genetic variation: maternal storage effect. Evolution 71:449–457

Yamamichi M, Letten AD (2020) Rapid evolution promotes fluctuation-dependent species coexistence. bioRxiv. https://doi.org/10.1101/2020.08.05.239053

Yamamichi M, Miner BE (2015) Indirect evolutionary rescue: prey adapts, predator avoids extinction. Evol Appl 8:787–795

Yamamichi M, Yoshida T, Sasaki A (2011) Comparing the effects of rapid evolution and phenotypic plasticity on predator-prey dynamics. Am Nat 178:287–304

Yamamichi M, Yoshida T, Sasaki A (2014) Timing and propagule size of invasion determine its success by a time-varying threshold of demographic regime shift. Ecology 95:2303–2315

Yamamichi M, Meunier CL, Peace A, Prater C, Rúa MA (2015) Rapid evolution of a consumer stoichiometric trait destabilizes consumer–producer dynamics. Oikos 124:960–969

Yamamichi M, Hairston NG Jr, Rees M, Ellner SP (2019a) Rapid evolution with generation overlap: the double-edged effect of dormancy. Theor Ecol 12:179–195

Yamamichi M, Klauschies T, Miner BE, van Velzen E (2019b) Modelling inducible defences in predator–prey interactions: assumptions and dynamical consequences of three distinct approaches. Ecol Lett 22:390–404

Yamamichi M, Lyberger K, Patel S (2019c) Antagonistic coevolution between multiple quantitative traits: matching dynamics can arise from difference interactions. Popul Ecol 61:362–370

Yamamichi M, Kyogoku D, Iritani R, Kobayashi K, Takahashi Y, Tsurui-Sato K, Yamawo A, Dobata S, Tsuji K, Kondoh M (2020) Intraspecific adaptation load: a mechanism for species coexistence. Trends Ecol Evol 35:897–907. https://doi.org/10.1016/j.tree.2020.05.011

Yoshida T, Jones LE, Ellner SP, Fussmann GF, Hairston NG Jr (2003) Rapid evolution drives ecological dynamics in a predator-prey system. Nature 424:303–306

Yoshida T, Hairston NG Jr, Ellner SP (2004) Evolutionary trade-off between defence against grazing and competitive ability in a simple unicellular alga, *Chlorella vulgaris*. Proc R Soc B Biol Sci 271:1947–1953

Yoshida T, Ellner SP, Jones LE, Bohannan BJM, Lenski RE, Hairston NG Jr (2007) Cryptic population dynamics: rapid evolution masks trophic interactions. PLoS Biol 5:1868–1879

Chapter 3
The Possibility that the Evolution of Sexual Traits Stabilizes Ecological Communities

Kazuya Kobayashi

Abstract Sexual reproduction is a common feature of eukaryotes which include dominant members in many biological communities, especially on the hot spots of biodiversity. Given its ubiquity, sex may have an important role to characterize the community structures. In this chapter, we explore the possibility that sex plays an important role to sustain the community complexities. First, we overview several properties of sex and various phenomena caused by it. Then, we demonstrate the effect of sex on the communities by introducing recent studies and discuss future directions.

Keywords Biological community · Evolution · Intraspecific competition · Density-dependence · Population dynamics · Sexual selection

3.1 Mechanisms for a Stabile Community

There are several definitions of stability in ecology (McCann 2000). To avoid confusion, here, stability is defined as follows: a state in which the population dynamics of all species constituting an ecosystem remain within a certain range. As long as this stability condition is satisfied, any species in the ecosystem is never extinct or overgrown, although the abundances of the species in the ecosystem will fluctuate within a range. This definition of stability does not allow for the extinction of any species in the community; thus, it does not include the coexistence of multiple species that is achieved by the balance between extinction and speciation. In addition, we do not consider models assuming migration from outside of the particular region. For example, an island population model that supposes immigration from continental regions without considering the population dynamics in the

K. Kobayashi (✉)
Hokkaido Forest Research Station, Field Science Education and Research Center, Kyoto University, Shibecha, Japan
e-mail: kobayashi.kazuya.3w@kyoto-u.ac.jp

A. Mougi (ed.), *Diversity of Functional Traits and Interactions*, Theoretical Biology,
https://doi.org/10.1007/978-981-15-7953-0_3

continental regions may satisfy the stability described above, but it is not considered here.

The remaining mechanism to achieve this stability is a negative density-dependent effect (NDDE) on population growth. To keep the abundance of a species in a certain range, the population growth rate needs to be reduced when the abundance moves towards the upper limit of the range, and to be increased when the abundance moves towards the lower limit of the range. Thus, a negative correlation is expected between the abundance and the population growth rate. This is the NDDE, which stabilizes the population dynamics of the species. If the NDDE is a general property for the species in a community, the community will also be stable. If such a negative correlation is absent, the abundance of the species changes stochastically, and thus, sometimes becomes zero. Then, the entire community will become unstable as the theoretical studies predicted (May 1972; Allesina and Tang 2012; Barabás et al. 2017).

Density-dependent population growth has been explained by two mechanisms: the predation pressure from natural enemies (Elton 1927; Murdoch 1969) and the intraspecific competition for resources (Gause 1934; Macarthur and Levins 1967). In the former mechanism, the NDDE is caused by the dynamics of predation pressure arising from the population dynamics of natural enemies, including predators, parasites and pathogens. The population dynamics of prey and predators are mutually controlled by each other. An increase in predator abundance reduces the prey abundance. A decrement in the prey abundance leads to a reduction in the predator abundance. After a decrease in predator abundance, the abundance of prey increases again. In the later mechanism relying on resource competition, it is assumed that each coexisting species uses a different niche space. Under this assumption, the competition for resources is stronger within a species than between species; thus, the population growth rate of a species depends mainly on the availability of resources and the conspecific abundance. In this case, the NDDE is caused by the intraspecific competition for resources, depending on the conspecific abundance.

These arguments about the mechanisms controlling population dynamics are classical theories in community ecology. Despite considerable efforts in this field, the elucidation of the mechanisms required to sustain biodiversity in nature is still one of the challenging tasks in ecology. Here, using the knowledge obtained from behavioural ecology or evolutionary ecology, I propose a hypothesis that sexual traits play an important role in sustaining biodiversity.

3.2 Evolutionary Optimum and Population Growth

As with the definition of stability, we need to clarify the definition of fitness because some community ecologists use the term differently than evolutionary ecologists (HilleRisLambers et al. 2012). In community ecology, fitness is occasionally used in the sense of the reproductive ability or number of offspring per individual; thus, it seems to be one of the demographic parameters for the species or population

(Hubbell 2001; Chave 2004; Bolnick et al. 2011). Fitness in evolutionary ecology is defined based on the relative genetic contribution towards the next generation within the population of a species. This difference occasionally causes a fatal error in the predictions of population dynamics because natural selection does not always favour an action for the good of the species. Particularly in the case where the amount of resources each individual acquires depends not only on the individual's own traits but also on those of the conspecific competitor for the resources, natural selection favours traits that increase the resource acquisition of an individual more than that of a competitor, even if the traits reduce the value of the resources. For the analysis of this situation, the evolutionary game theory is effective.

One of the most famous evolutionary games is the Hawk–Dove game (Maynard Smith and Price 1973), which supposes that two individuals competing for the resources behaves in one of two strategies, hawk or dove. Individuals with the hawk strategy are very aggressive and always fight for resources, while those with the dove strategy never fight for resources and always withdraw from them whenever they encounter an aggressive hawk. Thus, when the hawk and dove meet in front of resources, the hawk always obtains the whole resource. When hawks meet in front of resources, they fight, the winner takes all the resources, and the loser is injured. For simplicity, all hawks are assumed as equal in their competitive ability so that each hawk has an equal chance (50%) to be the winner in the competition. When doves meet in front of resources, they share the resources equally; thus, each dove always receives half of the resources. Therefore, under this condition, the fitness represented by the acquired resource amount of an individual depends not only on its own strategy but also on the strategy of the competitor. These conditions give the fitness function for each of the strategies as follows:

$$W_H = p(V/2 - C/2) + (1 - p)V \tag{3.1}$$

$$W_D = (1 - p)V/2 \tag{3.2}$$

where V is the value of the resources, C is the cost of injury, and p is the proportion of hawks in the population. In this model, the cost C caused by the competition between hawks reduces the total value of the resources at the population level. Thus, the population growth rate is maximized when all individuals adopt the dove strategy, and decreased by the increment of competition (interaction between hawks). However, a population that consists solely of doves is easily invaded by the hawk strategy because the hawk always obtains the entire resource without competition from doves. Thus, the dove strategy is evolutionarily unstable, and natural selection promotes the evolution of the hawk strategy, although it decreases the population growth rate of the species due to the competition between hawks.

The game theory shows that natural selection sometimes favours the trait that reduces the population growth rate of a species due to the intraspecific competitions between individuals. Next, focusing on mating competition, which is a fairly common reason to compete with conspecific individuals in nature, the effect of evolution on population dynamics is explored.

3.3 Evolution of Sex and Sex Allocation

We can observe numerous species using sexual reproduction in the natural world. This is mysterious because sexual reproduction requires numerous short-term costs that asexual organisms mostly avoid (Maynard Smith 1971; Williams 1975; Bell 1982; Lehtonen et al. 2012). One of the largest costs of sex is the resource allocation to males (Maynard Smith 1971; Dunbrack et al. 1995; Jokela et al. 1997; Innes et al. 2000; Wolinska and Lively 2008; Lehtonen et al. 2012; Kobayashi et al. 2013; Kobayashi and Hasegawa 2016). Although males cannot contribute to population growth rates, most sexual organisms allocate their resources to both sexes equally. The evolutionary reason for this equal allocation in sexual organisms is explained by the sex allocation theory (Fisher 1930), which is a pioneer study of evolutionary games.

Fisher's sex allocation theory explains the generality of the equal allocation of sexual organisms. If the number of males is larger than that of females in the population, the males can obtain less than one mate on average. Thus, the genetic contribution to the next generation per individual is higher in the females than the males, which promotes the evolution of a female-biased allocation in the male-biased population. In contrast, in populations where the number of females is larger than that of males, the males can obtain more than one mate on average. Therefore, the genetic contribution is higher in males than in females, which promotes the evolution of a male biased allocation in the female-biased population. Consequently, these evolutionary dynamics result in the evolution of an equal allocation in the population. Therefore, most sexual organisms allocate their resources almost equally to both sexes despite a reduction in population growth rates of the species compared to all female conditions.

The evolution of equal allocation requires random mating with large unstructured populations. If these conditions were violated, the evolution of biased allocations is predicted (Hardy 2002; West 2009). Supposing the situation where sons compete with each other for mating in a local sub-population, the allocation to sons (or the male function, such as pollen in plants) reduces the mating success rates for each son because strong mating competition among sons occurs for a limited number of mating partners, whereas the greater allocation to daughters creates more mating partners for sons. This situation is called local mate competition and was analysed by Hamilton (1967) using the framework of game theory. In this situation, the fitness of the mother W is

$$W = Bs^* + B(1 - s^*)\frac{s^* + (n-1)s}{1 - s^* + (n-1)(1-s)} \tag{3.3}$$

where B is an amount of resources for reproduction, n is a number of mothers within a local mating population, and s and s^* are allocated to the females of wild and mutant types, respectively. The first term is the fitness arising from the allocation to daughters (or seeds), and the second term arises from the allocation to sons (or pollen) that are affected by the ratio of females to males in the local mating population. To determine the evolutionarily stable allocation, the derivative of W is taken with respect to s^* and set $s^* = s$ and the value is set at zero:

$$\left.\frac{\partial W}{\partial s^*}\right|_{s^*=s} = \frac{n+1-2ns}{n(1-s)} = 0 \tag{3.4}$$

The solution is

$$s = \frac{n+1}{2n} \tag{3.5}$$

As n increases, the evolutionally stable allocation gradually approaches 0.5. If a single mother occurs in the local mating population ($n = 1$), this female should allocate most of her resources to daughters because the competition over mating occurs only between sons. In general, the competition among sons over mating promotes the evolution of extraordinary female-biased allocation depending on the number of mothers in the local mating population (Hamilton 1967).

The allocation to females is strictly linked with the population growth rates of the species because males cannot reproduce by themselves. Thus, the evolutionary dynamics of the sex allocation predicted by the local mate competition theory must provide a NDDE on the population growth (Hassell et al. 1983; Zhang and Jiang 1995; Kobayashi and Hasegawa 2016; Kobayashi 2017). In a species at a low density, where n will be small, sons must compete against each other for mating, which promotes the evolution of female-biased allocation (a reduction of the allocation to sons); thus, the population growth rate of a species with a low density increases. In species at high density, competition over mating occurs among unrelated individuals, which promotes the evolution of equal allocation; thus, the population growth rate of the species is reduced compared to that in the condition of female-biased allocations. The simulation models incorporating this mechanism demonstrated that hundreds of plant species can co-occur over 10,000 generations (Kobayashi 2017). This model supposes biological communities at a single trophic level under a homogeneous environment and a species-specific amount of available resources to pollen and seeds, which reflects differences in the species-level adaptation to the environment. Under this condition, the traditional models predict that all other species will be extinct, leaving the species that has the largest amount of available resources. However, in the model, hundreds of species firmly coexist because the common species with large amounts of available resources invested almost the same amounts in pollen and seeds, but rare species invest more in seeds. Thus, the evolutionary dynamics of sex allocation provide a buffer against species differences in the population growth and affect the structure of the ecological community by controlling the population dynamics of the species.

3.4 Evolution of Sexual Harassment

Another example is sexual harassment, which is a trait of one sex (usually the males) that increases its fertilization success rate but reduces the fecundity of its mating partners. Although many empirical studies have provided evidence of the density or frequency dependence of sexual selection (Andersson 1994; Levitan 2004; Gosden and Svensson 2009; Takahashi and Kawata 2013), the relationship between the density dependence of sexual selection and the population growth rate remains obscure (Kokko and Brooks 2003; Kokko and Rankin 2006). Recently, using the framework of game theory, I constructed a model of sexual harassment including dynamic feedback effects between the population dynamics and the evolutionary dynamics (Kobayashi 2018, 2020; Iritani 2020).

For species with high densities, sexual selection favours males with strong harassment to increase their fertilization success rates, even at the expense of the fecundity of their mating partners because a high density means that there are many competitors for mating. The evolution of such strong harassment results in the reduction of the population growth rate of the species. At low densities, which mean that there are few competitors, such harassment becomes disadvantageous for the harasser because it decreases the fecundity of mating partners; thus, a low density decreases the occurrence of harassment in the species so that the population growth rate of the species increases when the species abundance decreases. Supposing a self-compatible flowering plant species, this condition is modelled as follows (Kobayashi 2018, 2020; Iritani 2020):

$$W = \frac{h^*}{h^* + (n-1)h} nF \left[1 - \frac{h^* + (n-1)h}{n}\right] + F\left[1 - \frac{h^* - (n+1)h}{n}\right] \quad (3.6)$$

where F is the fecundity per individual (amount of seeds) without harassment, n is the number of individuals in the local population (pollen dispersal area), and h and h^* are the harassment strength of the wild and mutant types, respectively. The first term is the fitness as male, which is obtained by multiplication of the fertilization success rate for the individual (the proportion of the harassment of the mutant individual in the total harassment), the total fecundity of the local population without harassment and the total effect of harassment. The second term is the fitness as female, which represents the fecundity, including the effect of harassment. To determine the evolutionarily stable strategy for the harassment level, the derivative of W is taken with respect to h^*, $h^* = h$ and the value is zero:

$$\left.\frac{\partial W}{\partial h^*}\right|_{h^*=h} = \frac{n - 1 - nh - h}{nh} F = 0 \quad (3.7)$$

The solution is

$$h = \frac{n-1}{n+1} \tag{3.8}$$

Then, in the population with evolutionarily stable harassers, the realized fecundity per individual (R) becomes

$$R = \frac{2}{n+1}F \tag{3.9}$$

This equation shows that the reproductive rate per individual gradually nears zero as the density n increases. As shown by the sex allocation theory, the evolution of sexual harassment also reduces the population growth rates depending on the strength of the mating competition (the number of competitors). Thus, the model clearly shows that the evolution of sexual harassment can control the population growth rate according to the density and can stabilize the population size. Qualitatively, the same results were obtained from the models supposing self-incompatible and dioecious species (Kobayashi 2018).

As shown by the simulation of sex allocation, the simulation models incorporating sexual harassment also demonstrate that hundreds of plant species can co-occur over 10,000 generations despite the models not supposing niche differentiation and natural enemies (Kobayashi 2018). Notably, the NDDE arising from sexual harassment is stronger than that from sex allocation. In the case of sex allocation, the differences in the reproductive potentials between species are buffered up to two times because the evolutionally stable allocation gradually approaches 0.5 with an increasing density (n). In contrast, the evolution of sexual harassment infinitely buffers the interspecific differences in the reproductive potentials because the realized fecundity per individual approaches zero as n increases. These results suggest that sexual organisms have an intrinsic feature to regulate their abundance in general.

Theoretically, there is no limit to the number of co-existable species if the NDDE occurs in all of these species; however, in the simulations of sexual harassment and sex allocation that began with 1,000 species, some species were extinct (Kobayashi 2017, 2018). This is partly because these species failed to achieve evolution to increase their population growth before extinction due to their small population size. As recent eco-evolutionary studies pointed out (Pelletier et al. 2009; Post and Palkovacs 2009; Bolnick et al. 2011), mutation or genetic variation is essential to achieve evolution, but it will be difficult to secure them sufficiently for species with small population sizes. Therefore, under the simulated condition supposing a finite number of individuals, extinction occurs stochastically. Indeed, simulations showed that the number of species co-occurring over 10,000 generations was positively correlated with the total number of individuals in the community, which suggests that sustaining the intraspecific genetic diversity is important to avoid extinction in a species.

3.5 Rank-Abundance Diagram and the Frequency of Competitors

It is likely that a general theory of biodiversity exists because some of the measures of biodiversity often show similar patterns over a wide range of ecological communities. One example is the species abundance distribution that is generally drawn in the rank-abundance diagram (RAD), where the logarithm of abundance is plotted against the rank of the species in abundance (Hubbell 2001; Magurran 2004). It is known that the RAD of many ecological communities at a particular trophic level shows an S-shaped form.

The simulations including the NDDE of sex allocation (Kobayashi 2017) and sexual harassment (Kobayashi 2018) succeeded in reproducing the S-shaped form of RAD, which is the similar pattern observed in nature. In the study of the sex allocation model, I showed why the species abundance exhibited an S-shaped form in the model. The evolutionarily stable sex allocation is shown by Eq. (3.5): $s = (n + 1)/2n$. Replacing the number of individuals in a local population (n) with the expectation of a zero-truncated Poisson distribution, the relationship between the evolutionarily stable allocation and the density becomes S-shaped as follows:

$$s = \frac{\lambda + 1 - e^{-\lambda}}{2\lambda} \tag{3.10}$$

where s and λ indicate an evolutionarily stable allocation to females and the expected number of conspecific individuals in the pollen dispersal area (local population), respectively. In the diagram with the allocation toward females on the x-axis and the abundance on the y-axis, the sex allocation evolves towards female and male bias in the region under and above this equation line, respectively. When the evolution of the female-biased allocation makes the population growth rate of a species higher than that of the community average, the density of the species increases and crosses the line of Eq. (3.10) from the bottom to the top. Then, the evolution of the sex allocation changes its direction and begins to move towards a male bias. Similarly, the evolution of male bias leads to a decrease in the density of the species and crosses the line of Eq. (3.10) from the top to the bottom. Then, the evolution of female bias occurs again. As a result, the sex allocation and density of any species are kept close to the line of Eq. (3.10). Because the population growth rate is considered to be multiplied by the sex allocation and amount of investable resources, this evolutionary dynamic of sex allocation creates the S-shaped form in the RAD as long as the amount of resources available for each species is continuously distributed.

This analytical method cannot be applied to the sexual harassment model because the evolutionarily stable strategy of sexual harassment must consider another constraint when the density fluctuates. When the number of competitors suddenly increases, the fitness becomes negative because the harassment is too strong, which means the failure of reproduction. Thus, the probability distribution of the number of competitors should be included in the model as follows:

$$W = \sum_{n=1}^{\infty} W_n p_n \tag{3.11}$$

where W_n is the same with W in Eq. (3.11), and p_n is the probability that the number of competitors is n. Although the analytical solution of this equation remains obscure, the simulation models show that the NDDE arising from sexual harassment also reproduces the S-shaped form in the RAD. Since the relative population growth rates range from 0 to 1, the relationship between the density (the number of competitors in a local population) and the relative population growth rate will be similar to the inverse sigmoid function. This relationship between the density and population growth rate may be the reason why the S-shaped form is widely found in the RAD of field data.

The definition of competitors in this context should be noted. In both models (sexual harassment and sex allocation), mating competition is the key factor driving the evolutionary dynamics. When we adopt the biological species concept, which defines species as groups of interbreeding natural populations that are reproductively isolated from other such groups (Mayr 1970; Abbott et al. 2008; Rabosky 2016), mating competition must occur within the species. In other words, sexual organisms must compete for mates at high densities regardless of the available resources (or niche space). Thus, the NDDE arising from mating competition inevitably strengthens intraspecific competition over interspecific competition, which agrees with the necessary condition of coexistence predicted by the classic competitive exclusion principle (Gause 1934).

3.6 Future Directions

The study focusing on the relationship between sexual selection and population dynamics has just begun compared with the long history and sophistication of the sex allocation theory; thus, further studies are required to obtain experimentally testable predictions. In a previous study of sexual harassment (Kobayashi 2018), the evolution of female traits, such as tolerance to harassment and mate choice, was not considered. In the case of tolerance evolution, a co-evolution between harassment and tolerance will occur. However, the co-evolution will not alter the negative correlation between the population growth rate and density. In contrast, the evolution of female choice might change the evolutionary consequence of the model dramatically. The existence of female choice may promote the evolution of the male traits that are beneficial for females, such as nuptial gifts and parental care by males. These male traits will improve the population growth rate of the species. Thus, it is a challenging issue to predict population dynamics under the more broad types of sexual selection that promote the evolution of various sexual traits (Kobayashi 2019). To obtain some prediction that can be applied to actual organisms, it will be necessary to survey the options available for the organism, which will be constrained by the genetic basis for the traits. The real world in front of us will be built on these complex multilayer dynamics.

It would be difficult to think that the patterns found in this world that is abundant with sexual reproductive organisms are not influenced by sexual reproduction. The fact that sexual organisms flourish everywhere on the world suggests that the concept of sexual selection is important for understanding the various biological communities and ecosystem functions, as well as the niche concepts. Integrating multi-level dynamics, including genes, individuals, populations, species and communities, will open a new era of ecology. On the stage called as ecosystems, where the various actors are complexly intertwined, sex may be playing the leading role.

Acknowledgements I'd like to thank J. Yoshimura, E. Hasegawa, K. Tsuji, K. Matsuura, M. Kondoh, R. Tateno, T. Yoshioka, T. Nishida, K. Takakura, K. Kawatsu and D. Kyogoku for their helpful discussions. This work was partially supported by the Grants-in-Aid for Scientific Research of the Japan Society for the Promotion of Science (no. 17H05048).

References

Abbott RJ, Ritchie MG, Hollingsworth PM (2008) Introduction. Speciation in plants and animals: pattern and process. Philos Trans R Soc Lond Ser B Biol Sci 363:2965–2969. https://doi.org/10.1098/rstb.2008.0096

Allesina S, Tang S (2012) Stability criteria for complex ecosystems. Nature 483:205–208. https://doi.org/10.1038/nature10832

Andersson MB (1994) Sexual selection. Princeton University Press, Princeton

Barabás G, Michalska-Smith MJ, Allesina S (2017) Self-regulation and the stability of large ecological networks. Nat Ecol Evol 1:1870–1875. https://doi.org/10.1038/s41559-017-0357-6

Bell G (1982) The masterpiece of nature: the evolution and genetics of sexuality. Springer, Cham

Bolnick DI, Amarasekare P, Araújo MS et al (2011) Why intraspecific trait variation matters in community ecology. Trends Ecol Evol 26:183–192. https://doi.org/10.1016/j.tree.2011.01.009

Chave J (2004) Neutral theory and community ecology. Ecol Lett 7:241–253. https://doi.org/10.1111/j.1461-0248.2003.00566.x

Dunbrack RL, Coffin C, Howe R (1995) The cost of males and the paradox of sex: an experimental investigation of the short-term competitive advantages of evolution in sexual populations. Proc R Soc London B Biol Sci 262:45–49

Elton CS (1927) Animal ecology. University of Chicago Press, Chicago

Fisher RA (1930) The Genetical theory of natural selection. Clarendon, Oxford

Gause GF (1934) Experimental analysis of Vito Volterra's mathematical theory of the struggle for existence. Science 79:16–17

Gosden TP, Svensson EI (2009) Density-dependent male mating harassment, female resistance, and male mimicry. Am Nat 173:709–721. https://doi.org/10.1086/598491

Hamilton WD (1967) Extraordinary sex ratios. Science 156:477–488. https://doi.org/10.1126/science.156.3774.477

Hardy I (2002) Sex ratios: concepts and research methods. Cambridge University Press, Cambridge

Hassell MP, Waage JK, May RM (1983) Variable parasitoid sex ratios and their effect on host-parasitoid dynamics. J Anim Ecol 52:889. https://doi.org/10.2307/4462

HilleRisLambers J, Adler PB, Harpole WS et al (2012) Rethinking community assembly through the lens of coexistence theory. Annu Rev Ecol Evol Syst 43:227–248. https://doi.org/10.1146/annurev-ecolsys-110411-160411

Hubbell SP (2001) The unified neutral theory of biodiversity and biogeography. Princeton University Press, Princeton

Innes DJ, Fox CJ, Winsor GL (2000) Avoiding the cost of males in obligately asexual Daphnia pulex (Leydig). Proc R Soc London Ser B Biol Sci 267:991–997

Iritani R (2020) Gametophytic competition games among relatives: when does spatial structure select for facilitativeness or competitiveness in pollination? J Ecol 108:1–13. https://doi.org/10.1111/1365-2745.13282

Jokela J, Lively CM, Dybdahl MF, Fox JA (1997) Evidence for a cost of sex in the freshwater snail Potamopyrgus antipodarum. Ecology 78:452–460

Kobayashi K (2017) Sex allocation promotes the stable co-occurrence of competitive species. Sci Rep 7:43966. https://doi.org/10.1038/srep43966

Kobayashi K (2018) Sexual selection sustains biodiversity via producing negative density-dependent population growth. J Ecol 107:1433–1438. https://doi.org/10.1111/1365-2745.13088

Kobayashi K (2019) Sexual reproduction and diversity: connection between sexual selection and biological communities via population dynamics. Popul Ecol 61:135–140. https://doi.org/10.1002/1438-390X.1029

Kobayashi K (2020) Conditions for kin selection to bring cooperation and improve population growth: a response to Iritani. J Ecol 108:14–16. https://doi.org/10.1111/1365-2745.13281

Kobayashi K, Hasegawa E (2016) A female-biased sex ratio reduces the twofold cost of sex. Sci Rep 6:23982. https://doi.org/10.1038/srep23982

Kobayashi K, Yoshimura J, Hasegawa E (2013) Coexistence of sexual individuals and genetically isolated asexual counterparts in a thrips. Sci Rep 3:3286. https://doi.org/10.1038/srep03286

Kokko H, Brooks R (2003) Sexy to die for? Sexual selection and the risk of extinction. Ann Zool Fennici 40:207–219. https://doi.org/10.2307/23736526

Kobayashi (2019) https://doi.org/10.1111/1365-2745.13281

Kokko H, Rankin DJ (2006) Lonely hearts or sex in the city? Density-dependent effects in mating systems. Philos Trans R Soc Lond Ser B Biol Sci 361:319–334. https://doi.org/10.1098/rstb.2005.1784

Lehtonen J, Jennions MD, Kokko H (2012) The many costs of sex. Trends Ecol Evol 27:172–178

Levitan DR (2004) Density-dependent sexual selection in external fertilizers: variances in male and female fertilization success along the continuum from sperm limitation to sexual conflict in the sea urchin Strongylocentrotus franciscanus. Am Nat 164:298–309. https://doi.org/10.1086/423150

Macarthur R, Levins R (1967) The limiting similarity, convergence, and divergence of coexisting species. Am Nat 101:377–385. https://doi.org/10.1086/282505

Magurran AE (2004) Measuring biological diversity. Blackwell, Oxford. ISBN 0-632-05633-9

May RM (1972) Will a large complex system be stable? Nature 238:413–414. https://doi.org/10.1038/238413a0

Maynard Smith J (1971) The origin and maintenance of sex. In: Williams GC (ed) Group selection. Aldine Atherton Inc, Chicago, pp 163–175

Maynard Smith J, Price GR (1973) The logic of animal conflict. Nature 246:15–18

Mayr E (1970) Populations, species, and evolution: an abridgment of animal species and evolution. Harvard University Press, Harvard

McCann KS (2000) The diversity–stability debate. Nature 405:228–233. https://doi.org/10.1038/35012234

Murdoch WW (1969) Switching in general predators: experiments on predator specificity and stability of prey populations. Ecol Monogr 39:335–354. https://doi.org/10.2307/1942352

Pelletier F, Garant D, Hendry AP (2009) Eco-evolutionary dynamics. Philos Trans R Soc Lond Ser B Biol Sci 364:1483–1489. https://doi.org/10.1098/rstb.2009.0027

Post DM, Palkovacs EP (2009) Eco-evolutionary feedbacks in community and ecosystem ecology: interactions between the ecological theatre and the evolutionary play. Philos Trans R Soc Lond Ser B Biol Sci 364:1629–1640. https://doi.org/10.1098/rstb.2009.0012

Rabosky DL (2016) Reproductive isolation and the causes of speciation rate variation in nature. Biol J Linn Soc 118:13–25. https://doi.org/10.1111/bij.12703
Takahashi Y, Kawata M (2013) A comprehensive test for negative frequency-dependent selection. Popul Ecol 55:499–509. https://doi.org/10.1007/s10144-013-0372-7
West SA (2009) Sex allocation. Princeton University Press, Princeton
Williams GC (1975) Sex and evolution. Princeton University Press, Princeton
Wolinska J, Lively CM (2008) The cost of males in *Daphnia pulex*. Oikos 117:1637–1646
Zhang D-Y, Jiang X-H (1995) Local mate competition promotes coexistence of similar competitors. J Theor Biol 177:167–170. https://doi.org/10.1006/jtbi.1995.0235

Chapter 4
Spatiality and Community Stability

Akihiko Mougi

Abstract What allows interacting, diverse species to coexist in nature has been a central question in ecology, ever since the theoretical prediction that a complex community should be inherently unstable. Although the role of spatiality in species coexistence has been recognized, its application to more complex systems has been less explored. In this chapter, I argue that meta-community complexity, measured by the number of local food webs and their connectedness, elicits a self-regulating, negative-feedback mechanism and thus stabilizes food-web dynamics. Moreover, I argue that the presence of meta-community complexity can give rise to a positive food-web complexity-stability effect. Spatiality may play a more important role in stabilizing dynamics of complex, real food webs than expected from ecological theory based on the models of simpler food webs.

Keywords Meta-community · Migration · Food-web · Complexity · Stability

The world in which we live is undoubtedly spatial and species interactions are spatially limited. This spatiality has long been recognized as an essential property of ecology. For example, in 1958, Huffaker (1958) showed that a phytophagous mite and its predator did not persist in small, homogeneous habitats but could survive in more complex habitats that were comprised of distinct local habitats linked by low levels of dispersal. Thus, even if sub-populations or subcommunities are locally unstable, migrations can stabilize the entire system at a global spatial scale. This meta-community concept has played a key role in ecology, from the dynamics of individual species to the structure, dynamics, diversity, and stability of communities.

In this chapter, I argue that the meta-community concept is key to resolving two classical ecological paradoxes around the relationship between complexity and both stability and predictability. Complexity–stability paradox is a central issue even in

A. Mougi
Faculty of Life and Environmental Sciences, Shimane University, Matsue, Shimane, Japan

A. Mougi (ed.), *Diversity of Functional Traits and Interactions*, Theoretical Biology,
https://doi.org/10.1007/978-981-15-7953-0_4

current ecology that is based on the contradiction between May's theory (May 1972), which predicts a negative complexity–stability relationship whereby larger systems with diverse species and interactions are less stable than smaller systems, and earlier ecological views and observations (Elton 1958; MacArthur 1955). Here, I propose that the definition of complexity is important for solving this puzzle. May (May 1972) defined complexity in terms of the community or network complexity, considering the numbers of species nodes and interaction links (connectance). By contrast, Elton (1958) argued that ecosystem complexity includes not only species diversity but also spatial complexity, such as the spatial scale and heterogeneity, and pointed out that simplified ecosystems, such as agricultural or degraded ecosystems, are more prone to population oscillations than natural, complex ecosystems. This argument clearly relates to ecosystem size, including the spatial scale and community size, so I show that the simple extension of May's community complexity to Elton's ecosystem complexity leads to a positive complexity–stability relationship appearing, resolving this long-standing paradox.

The ability to predict an ecosystem's response to environmental disturbance or artificial control is another challenging task in ecology. The complexity–predictability relationship was originally examined by Yodzis (1988), who tackled this problem using a mathematical food-web model and showed that it was quite difficult to even make a qualitative prediction of species' responses to a perturbation. For example, an increase in the abundance of predators does not necessarily reduce the abundance of prey, contrary to the expected direct effect of a predator on prey, due to the many indirect pathways through which perturbations can be dampened or magnified masking or even counteracting this direct effect. This complexity within ecosystems combined with the low resolution at which interaction strengths can be quantified limit our ability to understand community dynamics because inherent estimation errors in interaction strengths can easily reverse the direction of species' responses to a perturbation. However, contrary to these theoretical predictions, natural ecosystems often appear to show expected responses (Strong 1992; Schoener 1993; Polis and Strong 1996; Estes et al. 2011); but see (Menge et al. 1994; Power et al. 1996)—for example, an increase in the abundance of sea otters reduces the abundance of sea urchins, whereas a decrease in the abundance of wolves increases the abundance of elk—and these expected patterns are widely consistent across a range of regions and communities. This contradiction between theory and natural observations suggests that ecosystems possess some mechanism that allows them to avoid the distractions of indirect effects. In this chapter, I argue that the meta-community complexity is also key to solving this puzzle.

4.1 Meta-Community Complexity and Community Stability

4.1.1 Spatiality Can Resolve the Complexity–Stability Paradox

The stabilizing effects of spatiality on food-web dynamics have been well studied. However, early studies only considered simple communities consisting of a few species (Tilman and Kareiva 1997; Amarasekare 2008), did not account for spatial effects (McCann et al. 2005; Gravel et al. 2011), did not examine interactions between food-web complexity and spatiality (Plitzko and Drossel 2015), or only recognized food-web complexity as a result of the spatial aggregation of local, simple sub-webs (Pillai et al. 2011). Thus, the importance of spatiality for the persistence of more complex food webs remained unclear.

Mougi and Kondoh (2016) considered a meta-community in which organisms randomly moved between numerous coexisting local food webs (Fig. 4.1). This "meta-food web" could be viewed as a network of networks and thus was characterized by two different kinds of network complexity: food-web complexity and meta-community complexity. Food-web complexity could be quantified by the number of species that were involved in a local food web, *N*, and the probability that a pair of species was connected by a trophic link, *C*. Similarly, meta-community complexity could be quantified by the number of local food webs, H_N, and the proportion of food-web pairs between which an organism could move, H_C (the proportion of realized links to all possible links among local food webs). Random networks were assumed for both the food-web and habitat structures. The strength of species migration between local food-web patches was given as *M*. By controlling

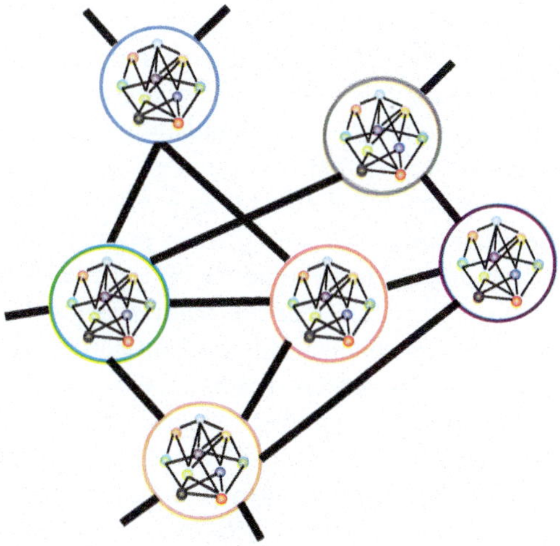

Fig 4.1 Schematic representation of a meta-community. Each large circle indicates a patch, while the different-colored small circles within the large circles indicate local populations of each species. Thin black links indicate interactions between local populations of different species and bold black links indicate species migrations between habitats

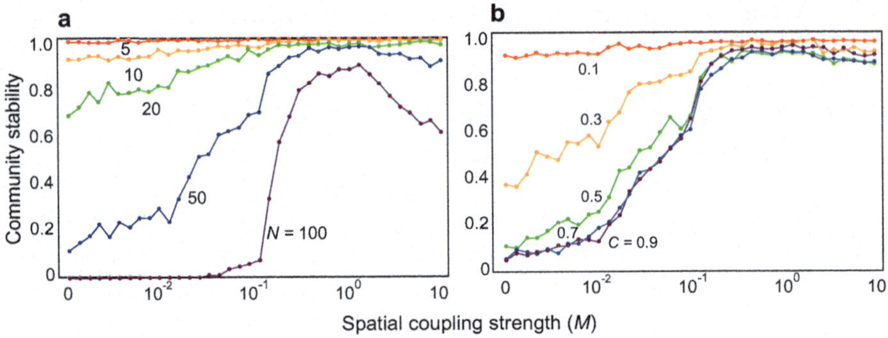

Fig. 4.2 Relationship between the spatial coupling strength (M) and stability under different scenarios. (**a**) Effect of species richness (N) assuming that connectance (C) = 0.5. (**b**) Effect of C assuming that N = 50. Different colors represent different levels of species richness and connectance. Habitat number (H_N) = 2, habitat connectivity (H_C) = 1. Taken from Mougi and Kondoh (2016)

the two network complexities (N, C, H_N, H_C) and M systematically, the role of space in the complexity–stability debate was revealed.

First, consider the simplest case of two local food webs ($H_N = 2$, $H_C = 1$). When the local food webs are isolated ($M = 0$), increasing the food-web complexity tends to destabilize the population dynamics (Fig. 4.2), which is consistent with May's prediction (May 1972). By contrast, when the two local food webs are coupled by migration ($M > 0$), the stability of the complex food web increases to that of a simple food web, revealing the stabilizing force of spatial complexity. However, when the coupling is very strong, stabilization does not arise because the whole meta-food web behaves as a single food web. This sharp unimodal response to increased coupling strength tends to appear in larger food webs and spatial stabilization can occur even when May's random network (May 1972) is assumed instead of a food web (Gravel et al. 2016).

The observation that migration has a stabilizing effect as long as within-species correlations among habitats are not too strong can be explained by the fact that (1) different communities may have different equilibria and (2) immigration from a high- to a low-density population can act as a self-regulation mechanism. This suggests that spatial heterogeneity in species composition among local food webs is necessary for the stabilizing effect to occur because no migration will occur between local food webs if their population densities are too similar. This observation is consistent with the widely accepted hypothesis that spatial heterogeneity can enhance the coexistence of many species (Levin 1976). Interestingly, however, the proposed mechanism differs from earlier hypotheses as it states, for instance, that spatiality allows multi-species coexistence via spatial segregation or niche differentiation, or that different stages of "succession" are maintained in the meta-community.

Next, consider a case with more than three local food webs. Here, meta-community complexity (H_N, H_C) can affect the system stability in two interesting

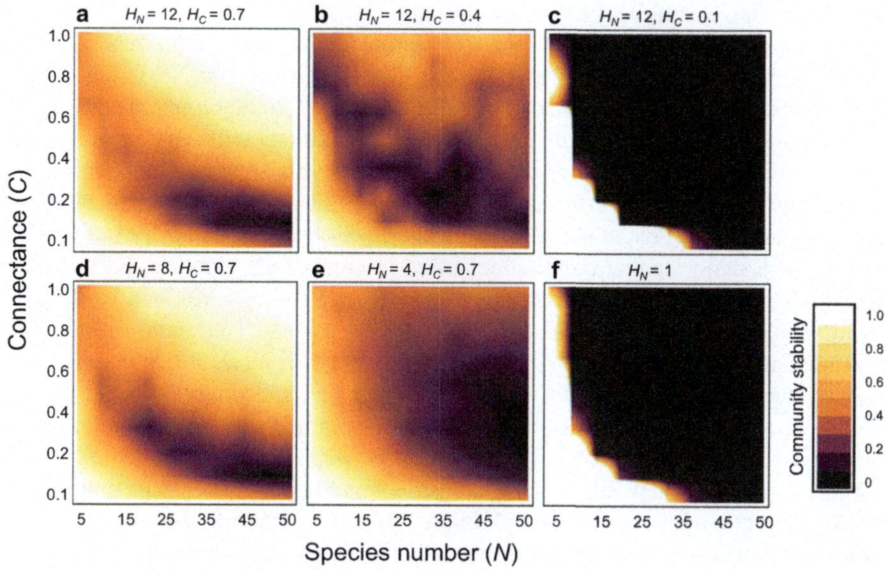

Fig. 4.3 Relationship between spatial complexity (habitat number H_N and habitat connectivity H_C) and stability with varying spatial coupling strength (M). Species richness (N) = 20, connectance (C) = 0.5. Taken from Mougi and Kondoh (2016)

ways: by stabilizing the community dynamics and by reversing the otherwise negative complexity–stability relationship of food webs to produce a positive relationship under an intermediated coupling strength (M) (Mougi and Kondoh 2016) (Fig. 4.3). This positive complexity–stability relationship is more prominent in a community that has more local food webs and connected pairs, suggesting that metacommunity complexity elicits the stabilizing effect of food-web complexity.

4.1.2 *Spatial Compartmentation Contributes to Community Stability*

May (1972) suggested that a non-random or compartmentalized network structure in which the subsystems have minimal linkages to other subsystems could stabilize complex systems. This was based on the expectations that isolated, small systems will be stable and that the existence of few links between them will buffer any ecological perturbations to the whole system (May 1972; Gardner and Ashby 1970; McNaughton 1978; Roberts and Tregonning 1980; Teng and McCann 2004). However, contrary to such expectation, several theoretical studies have shown that strongly compartmented food webs are in fact unstable (Pimm 1979; McCann 2011) or show moderate stability only under limited conditions (Grilli et al. 2016). Furthermore, little information is available regarding compartmentation within food-web data, except between major spatially distinct habitats, such as pelagic and benthic, shore and offshore, and land and marine environments (Pimm and Lawton 1980; Winemiller 1990; Polis and Strong 1996; Krause et al. 2003; Melián and Bascompte 2002); but see (Yodzis 1982; Rezende et al. 2009; Guimerà et al. 2010).

Undeniably, major spatial distinctions between habitats will separate an ecological community, creating compartments among spatially distinct subcommunities (Pimm and Lawton 1980; Polis and Strong 1996; Polis and Winemiller 1996; Huxel and McCann 1998; Post et al. 2000). However, very few studies have examined the role of such major spatial compartmentation in community stability. Furthermore, space has not been explicitly considered in previous studies (McCann et al. 2005; Gardner and Ashby 1970; McNaughton 1978; Roberts and Tregonning 1980; Teng and McCann 2004; Pimm 1979; McCann 2011; Grilli et al. 2016; Stouffer and Bascompte 2011), meaning that some species that link subcommunities are considered to be always interacting globally. By contrast, in real systems, subcommunities would be locally connected by interactions between local populations of some species and such local interactions across a boundary between subcommunities could propagate each community through migration. This meta-community view provides a different perspective on compartmentation, with subcommunities being coupled by not only species interactions but also spatial interactions through migration. However, it remains unclear how spatial compartments among spatially distinct subcommunities affect the maintenance of the whole community.

To examine this, I extended the meta-food-web model discussed above to examine the effects of spatial compartments on community stability (Mougi 2018) (Fig. 4.4). In the previous section, I assumed that local populations of each species interacted and moved among habitats (Fig. 4.4f). This meta-community view should also be true at a larger spatial scale. For example, consider two major spatially distinct habitats and a boundary habitat between them. Each organism can randomly move between these habitats, with the proportion of migratory species being controlled by a parameter p. When p is small, sub-food webs are linked by few interactions between local populations, being completely divided when $p = 0$ (Fig. 4.4a). By contrast, when p is large, sub-food webs are linked by many interactions between local populations (Fig. 4.4c). Similarly, when $M = 0$, the sub-food webs are isolated, whereas when M is extremely large, the sub-food webs are strongly coupled, behaving as a single food web.

In nature, distinct sub-food webs are linked by species interactions between local populations within a local habitat. This linkage can occur in two ways, depending on the degree of penetration of migratory species into external subsystems (the penetration degree). If penetration is limited to a local (boundary) habitat (boundary-separated subsystems), the sub-food webs of the main habitats will be locally and partially merged by interactions between species migrating between each sub-food web (Fig. 4.4b, c). By contrast, if the sub-food webs are linked across all of the habitats (globally connected subsystems), some species in a sub-food web may pass through the boundary habitat and appear in another sub-food web (Fig. 4.4e, f).

Here, I define "compartmentation" as the degree of spatial segregation between sub-food webs. The degree of compartmentation is controlled by three elements: the

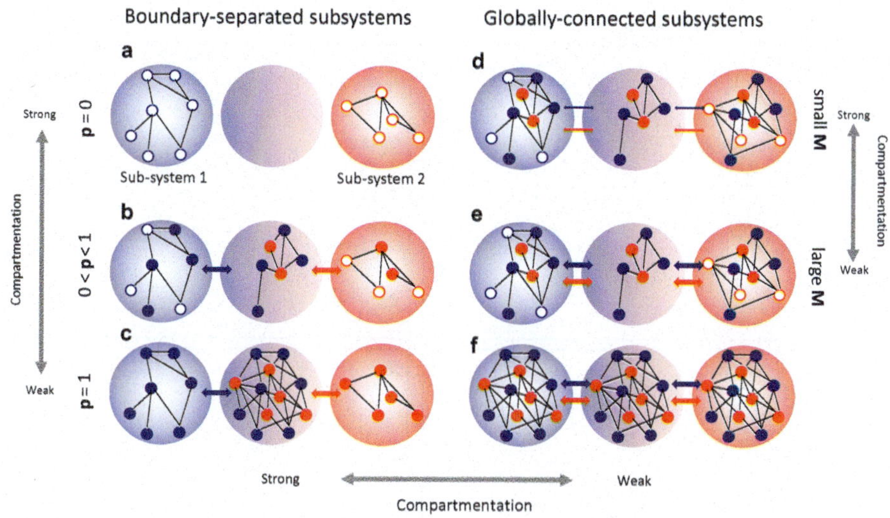

Fig. 4.4 Schematic representation of a spatially compartmented food-web model. (**a–c**) Boundary-separated subsystems. (**d–f**) Globally connected subsystems. When the proportion of migratory species (p) = 0, the subsystems are completely isolated, whereas when $p = 1$, the subsystems are completely merged. The large blue and red circles on the left and right, respectively, indicate patches and are the main habitats of the species represented by small blue and red circles, respectively. The medium circle is a boundary habitat where the two subsystems are locally and partially merged by interactions between species migrating from each main habitat. The small blue and red filled circles in each patch indicate local populations of migratory species in each main habitat, whereas the small blue and red open circles represent non-migratory species. The arrows between the patches indicate the degree of species migration—for example, the spatial coupling strength is lower in (**d**) than in (**e**). Taken from Mougi and Kondoh (2016)

proportion of migratory species (p), the spatial coupling strength (M), and the penetration degree (boundary-separated or globally connected subsystem). When p, M, and/or the penetration degree are low, compartmentation is expected to be strong (Fig. 4.4). Therefore, all three parameters can be controlled to examine the effects of spatial compartmentation on stability.

Consider a perfectly compartmented food web comprised of N species (each distinct sub-food web is comprised of $N/2$ species), any pair of which are connected to each other with probability C (connectance), which is defined as the proportion of realized interaction links to the maximum possible number of interaction links in a given network model in which no species move between habitats ($p = 0$; Fig. 4.5a). In this extreme case, it is trivial that the strength of spatial coupling, M, does not affect community stability (Fig. 4.5a, b). However, if the sub-food webs are coupled by migratory species, M dramatically alters the effects of p on community stability, with less connected food webs (smaller p) having higher stability when the habitats are weakly coupled (smaller M) and more connected food webs (larger p) having higher stability when the habitats are tightly coupled (larger M) (Fig. 4.5a, b). Further, more spatially compartmented food webs (smaller p and M) have lower

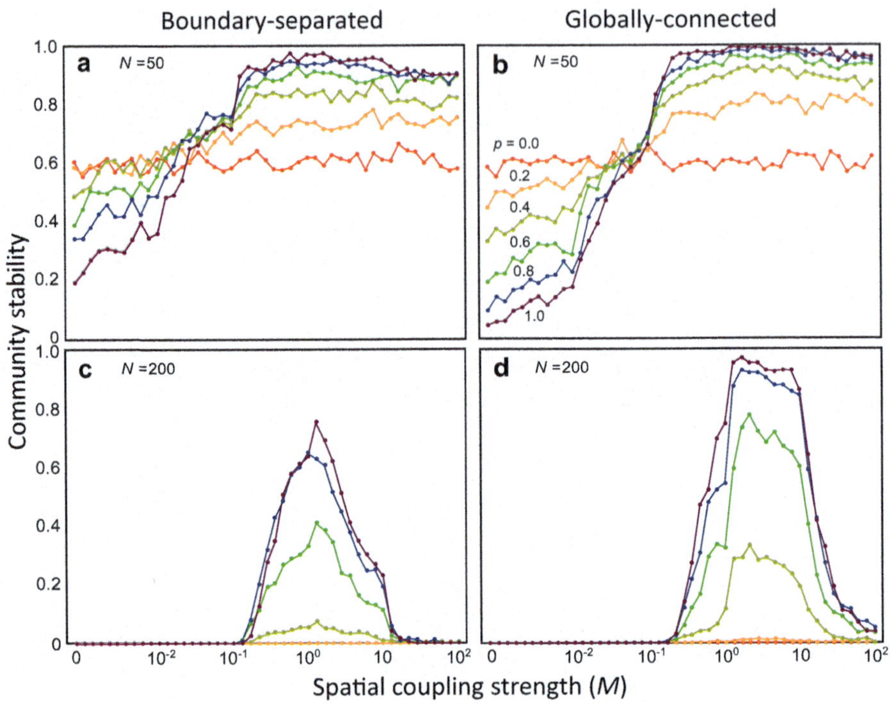

Fig. 4.5 Relationships between the spatial coupling strength and stability with varying proportions of migratory species where (**a**, **b**) $N = 50$ and (**c**, **d**) $N = 200$. (**a**, **c**) Boundary-separated subsystems. (**b**, **d**) Globally connected subsystems. Colors indicate different proportions of migratory species (p). Connectance (C) = 0.5. Taken from Mougi and Kondoh (2016)

stability than less spatially compartmented food webs (larger p and M) regardless of the penetration degree.

System size and species richness can greatly affect these results in four different ways. First, if the system becomes extremely large and M is large or small, stability can be extremely low or even zero regardless of the size of p (Fig. 4.5c, d). Such a system can be stable within a moderate range of M, however. [Note that destabilization is not likely to occur when M is large in smaller systems (Fig. 4.5a, b)]. Second, when M is moderate, the degree of compartmentation can dramatically affect stability, with less compartmented food webs tending to show high stability (Fig. 4.5c, d). Third, food webs with higher (smaller p and M) and lower (larger p and M) levels of spatial compartmentation have extremely low stability. Finally, the difference in stability between boundary-separated and globally connected subsystems becomes greater as systems become larger, with globally connected subsystems tending to be more stable than boundary-separated subsystems (Fig. 4.5c, d).

These results suggest that in smaller systems, less spatially compartmented food webs in terms of all three elements of compartmentation are likely to be more stable than more compartmented food webs. However, in larger systems, the degree of compartmentation becomes even more central to stability, with stabilization only arising under a moderate coupling strength, more spatially compartmented food webs not becoming stabilized, and stability tending to be higher in less compartmented food webs. Together, these results indicate that spatial compartmentation has a lesser and more negative effect on community stability in larger systems than in smaller ones, supporting the little evidence that is available on compartmented subcommunities within small systems (Pimm and Lawton 1980). Furthermore, it appears that large systems with major habitat divisions may be maintained by moderate levels of species migration rather than limited interactions between subcommunities, so it may not be spatial compartmentation that provides stabilization in these systems but rather the spatial limitation of species interactions. This suggests that it is the spatial limitation of species interactions rather than community interactions that plays a key role in maintaining communities.

May (1972) speculated that inherently stable, local, species-poor systems contribute to the stability of the whole system. However, the present model predicts that it is in fact species-rich local communities that contribute to the stability of the whole community, with the least stable communities with the greatest species diversity being most affected by immigration (Gravel et al. 2011; Mougi and Kondoh 2016). Thus, it appears that the effects of stabilizing self-regulation through migration (Mougi and Kondoh 2016) outweigh the stabilizing effects of species-poor systems.

4.2 Meta-Community Complexity and Predictability

Another important question is whether meta-community complexity affects an ecosystem's response to environmental disturbances. As explained in the Introduction, Yodzis (1988) theoretically showed that it is inherently difficult to predict the responses of species to press perturbation due to the indirect pathways that exist within a community. In the absence of meta-community complexity, many indirect pathways can either dampen or magnify the effects of perturbations, masking or even counteracting the expected direct effect of an interacting species on the perturbed species and resulting in low predictability of species' responses to perturbation. However, contrary to this theory, natural ecosystems often appear to show expected responses, suggesting that they possess some mechanism that allows them to avoid distractions from indirect effects.

Recent theories have provided some suggestions for improving predictability. For example, obtaining precise information on parameters related to interaction strengths allows accurate predictions to be made (Dambacher et al. 2002; Novak et al. 2011; Aufderheide et al. 2013). However, such predictions are practically impossible in real, complex ecosystems because errors will be associated with the estimations of many parameters, quickly reducing the accuracy of the predictions (Novak et al. 2011). Here, I show that spatiality alone, as a general and inherent feature of any natural ecosystem, can increase predictability.

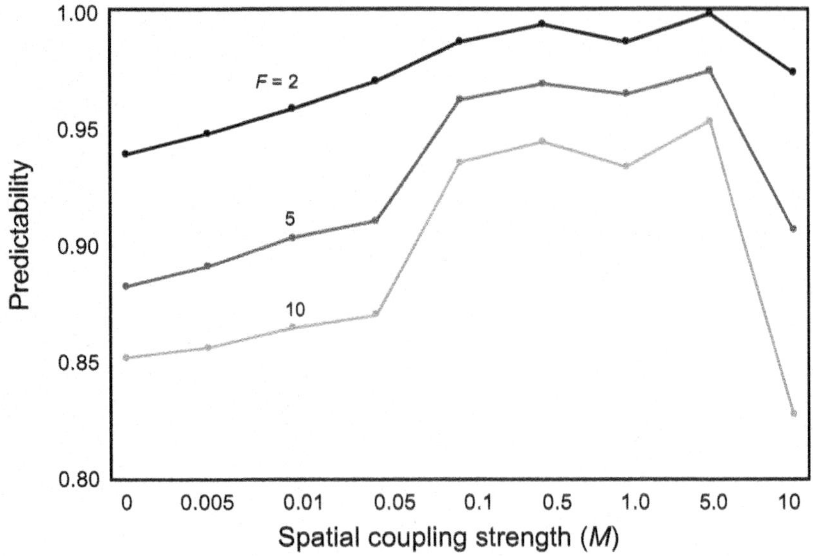

Fig. 4.6 Effects of estimation error (F) on predictability with varying spatial coupling strength. There were unimodal relationships between predictability and spatial coupling strength. Species richness (N) = 20, connectance (C) = 0.15, habitat number (H_N) = 12, habitat connectivity (H_C) = 0.6. Taken from Mougi (2017)

Consider the meta-food web example described in the previous section. By applying press perturbations and examining the long-term effects of small disturbances on species abundances (e.g., the effects of fisheries harvesting, pollution, eutrophication, or a controlled experiment), it is possible to determine the net effects (effects of all direct and indirect interactions) of such disturbances on the equilibrium species abundances. Predictability is evaluated based on the proportion of true predator (or prey) net responses that match the sign of the net effects after introducing an error to the interaction strength estimates. Therefore, to reveal the effects of spatiality on predictability, I calculated the mean proportion of net responses that matched the signs of the true net responses over many random iterations for each given network and error level, and controlled two measures of meta-community complexity (H_N and H_C) and M systematically (Mougi 2017).

I found that predictability in the focal local food webs is not high when they are isolated (M = 0), particularly when the estimation error is large (Fig. 4.6), which supports earlier theoretical studies (MacArthur 1955). However, when migration occurs ($M > 0$), predictability increases to reach a peak at a moderate coupling strength. I also found that predictability depends on spatial complexity at intermediate levels of M, tending to increase with an increasing number of local food webs (H_N) or connection probability (H_C) (Fig. 4.7). However, when the coupling between local food webs is too weak (smaller M) or too strong (larger M), predictability shows no clear relationship with spatial complexity. This suggests that high predictability arising from local food-web coupling is stronger in spatially complex

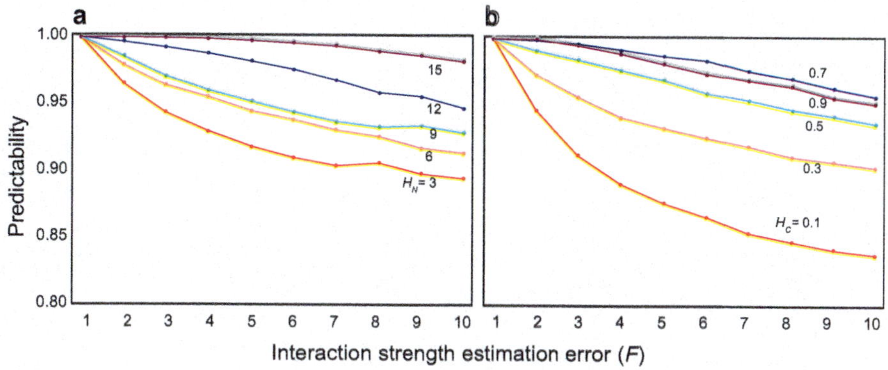

Fig. 4.7 Relationships between estimation error (F) and predictability with varying spatial complexity. (**a**) Effect of habitat number (H_N) where habitat connectivity (H_C) = 0.6. (**b**) Effect of H_C where H_N = 12. Colors indicate different levels of habitat number and connectivity. Species richness (N) = 20, connectance (C) = 0.15, spatial coupling strength (M) = 1. Taken from Mougi (2017)

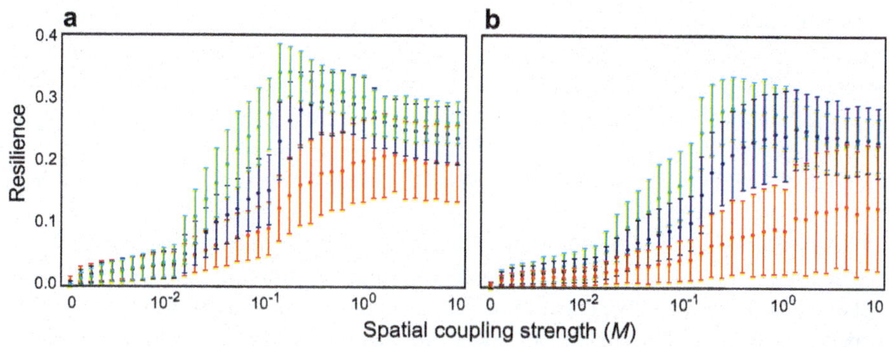

Fig. 4.8 Relationships between spatial coupling strength (M) and resilience with varying spatial complexity. (**a**) Effect of habitat number (H_N) where habitat connectivity (H_C) = 0.6. (**b**) Effect of H_C where H_N = 8. The resilience of the system (its capacity to return to an equilibrium after a short and small disturbance) was calculated as the absolute value of the greatest real part of the eigenvalues of the Jacobian matrix. Red, blue, and green indicate different levels of habitat number in (**a**) (H_N = 5, 10, and 20, respectively) and habitat connectivity in (**b**) (H_C = 0.3, 0.5, and 0.8, respectively). Taken from Mougi (2017)

ecosystems but that predictability will not arise when the coupling is so strong that the entire meta-food web behaves as a single food web.

The improvement of predictability with spatiality suggests that spatial networks and species migration have a buffering effect on perturbations via interaction pathways. The resilience of a system (its capacity to return to an equilibrium after a short and small disturbance) tends to be high when the predictability is high (Fig. 4.8). However, it remains unclear what causes this increased resilience. I hypothesized that spatiality weakens the magnitudes of indirect effects, allowing

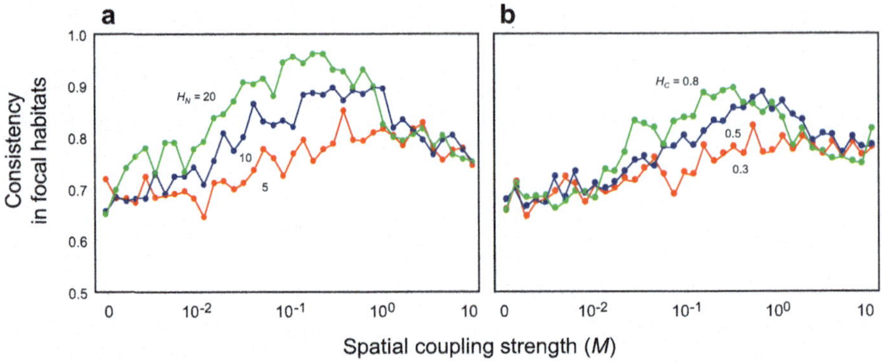

Fig. 4.9 Relationships between spatial coupling strength (M) and consistency with varying spatial complexity. (**a**) Effect of habitat number (H_N) where habitat connectivity (H_C) = 0.6. (**b**) Effect of H_C where $H_N = 8$. Different colors indicate different levels of habitat number and connectivity. Species richness (N) = 20, connectance (C) = 0.5. Taken from Mougi (2017)

food webs to be highly resilient. To test this hypothesis, I examined the relationship between spatiality and the degree of coincidence in directions between direct and net species' responses to perturbation (consistency), based on the idea that net species' responses to perturbation should show more consistent directions than those expected from direct responses if indirect effects are weakened by spatiality. As can be seen in Fig. 4.9, this expected pattern was observed. Thus, consistency is likely to be high under a moderate coupling strength and tends to increase with spatial complexity.

The increase in consistency due to spatiality can be explained by the buffering effect caused by the emigration of species to other habitats. In isolated food webs, net species' responses to perturbation are inherently reversed from what would be expected from direct responses due to many indirect effects masking or counteracting the direct effect (lower M in Fig. 4.10). However, if a local food web is connected to other local food webs, the perturbation can propagate not only to indirect pathways within the focal food web but also to other habitats through species emigration. Thus, by allowing individuals to move, the effects of perturbation can attenuate through indirect pathways, resulting in the species' net responses reflecting the direct effects (moderate M in Fig. 4.10). Finally, if local food webs are tightly coupled, perturbation strongly propagates to other habitats and easily returns to the local food web, resulting in a reduced difference in consistencies between focal and non-focal local food webs (Fig. 4.9a) and thereby intensifying the indirect effects and masking or counteracting the direct effects (higher M in Fig. 4.10).

Previous studies have shown that increasing network complexity can increase the predictabilities of interaction strengths or species' responses to sustained press (Iles and Novak 2016) or pulse (Berlow et al. 2009; Brose et al. 2005) perturbations using realistic, allometrically constrained parameterization and a non-linear functional response in food-web models. These common results are expected to originate from some mechanism that causes interaction strengths to become weaker as the

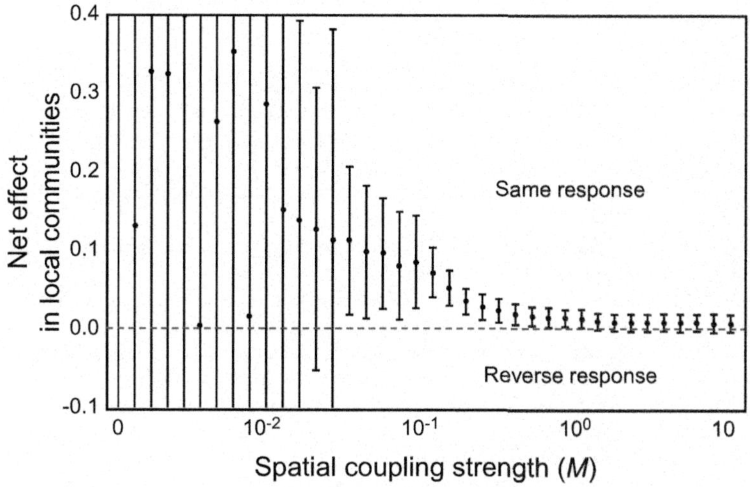

Fig. 4.10 Relationship between the spatial coupling strength (M) and the net effects in local communities. Black points and lines represent the mean values and error bars, respectively. The full error bars, including the lower values of M, are not shown. The red dashed line indicates the zero line, with values above this line indicating a positive response to perturbation (the directions of net and direct response are the same). Parameters are the same as in Fig 4.9a. Taken from Mougi (2017)

network size increases due to interaction effort allocations, where weaker interactions reduce propagations in indirect pathways (Schoener 1993; Menge et al. 1994). Contrary to earlier theories, the present theory does not rely on the assumption of interction strengths—rather, spatiality alone, as a general and inherent feature of any natural ecosystem, may allow high predictability.

4.3 Conclusion

In this chapter, I demonstrate that spatiality can have a positive effect on ecosystem stability and increase the predictability of ecosystem responses to perturbations, suggesting that there is a positive relationship between stability and predictability. In the presence of meta-community complexity, the effects of perturbation can spread to linked habitats through emigration, weakening the effects within local habitats. This mechanism would not only buffer the effects of the perturbation but also weaken the indirect effects, allowing high stability and predictability. Hence, habitat destruction and modification could make ecosystems uncontrollable.

References

Amarasekare P (2008) Spatial dynamics of foodwebs. Annu Rev Ecol Evol Syst 39:479–500
Aufderheide H, Rudolf L, Gross T, Lafferty KD (2013) How to predict community responses to perturbations in the face of imperfect knowledge and network complexity. Proc R Soc B 280:20132355
Berlow EL et al (2009) Simple prediction of interaction strengths in complex food webs. Proc Natl Acad Sci U S A 106:187–191
Brose U, Berlow EL, Martinez ND (2005) Scaling up keystone effects from simple to complex ecological networks. Ecol Lett 8:1317–1325
Dambacher JM, Li HW, Rossignol PA (2002) Relevance of community structure in assessing indeterminancy of ecological predictions. Ecology 83:1372–1385
Elton CS (1958) Ecology of invasions by animals and plants. Chapman & Hall, London
Estes JA et al (2011) Trophic downgrading of planet earth. Science 333:301–306
Gardner MR, Ashby WR (1970) Connectance of large dynamic (cybernetic) systems: critical values of stability. Nature 228:784
Gravel D, Canard E, Guichard F, Mouquet N (2011) Persistence increases with diversity and connectance in trophic metacommunities. PLoS One 6:e19374
Gravel D, Massol F, Leibold MA (2016) Stability and complexity in model meta-ecosystems. Nat Commun 7:12457
Grilli J, Rogers T, Allesina S (2016) Modularity and stability in ecological communities. Nat Commun 7:12031
Guimerà R, Stouffer DB, Sales-Pardo M, Leicht EA, Newman ME (2010) Origin of compartmentalization in food webs. Ecology 91:2941–2951
Huffaker CB (1958) Experimental studies on predation: dispersion factors and predator–prey oscillations. Hilgardia 27:343–383
Huxel GR, McCann K (1998) Food web stability: the influence of trophic flows across habitats. Am Nat 152:460–469
Iles AC, Novak M (2016) Complexity increases predictability in allometrically constrained food webs. Am Nat 188:87–98
Krause AE, Frank KA, Mason DM, Ulanowicz RE, Taylor WW (2003) Compartments revealed in food-web structure. Nature 426:282–285
Levin SA (1976) Population dynamic models in heterogeneous environments. Annu Rev Ecol Evol Syst 7:287–310
MacArthur R (1955) Fluctuations of animal populations, and a measure of community stability. Ecology 36:533–536
May RM (1972) Will a large complex system be stable? Nature 238:413–414
McCann KS (2011) Food webs. Princeton University Press, Princeton
McCann KS, Rasmussen JB, Umbanhowar J (2005) The dynamics of spatially coupled food webs. Ecol Lett 8:513–523
McNaughton SJ (1978) Stability and diversity of ecological communities. Nature 274:251–253
Melián CJ, Bascompte J (2002) Complex networks: two ways to be robust? Ecol Lett 5:705–708
Menge BA et al (1994) The keystone species concept: variation in interaction strength in a rocky intertidal habitat. Ecol Monogr 64:249–286
Mougi A (2017) Spatial complexity enhances predictability in food webs. Sci Rep 7:43440
Mougi A (2018) Spatial compartmentation and food web stability. Sci Rep 8:16237
Mougi A, Kondoh M (2016) Food-web complexity, meta-community complexity and community stability. Sci Rep 6:24478
Novak M et al (2011) Predicting community responses to perturbations in the face of imperfect knowledge and network complexity. Ecology 92:836–846
Pillai P, Gonzalez A, Loreau M (2011) Metacommunity theory explains the emergence of food web complexity. PNAS 108:19293–19298
Pimm SL (1979) The structure of food webs. Theor Popul Biol 16:144–158

Pimm SL, Lawton JH (1980) Are food webs divided into compartments? J Anim Ecol 49:879–898
Plitzko SJ, Drossel B (2015) The effect of dispersal between patches on the stability of large trophic food webs. Theor Ecol 8:233–244
Polis GA, Strong DR (1996) Food web complexity and community dynamics. Am Nat 147:813–846
Polis GA, Winemiller KO (1996) Food webs: integration of patterns and dynamics. Chapman & Hall, New York
Post DM, Conners ME, Goldberg DS (2000) Prey preference by a top predator and the stability of linked food chains. Ecology 81:80–14
Power ME et al (1996) Challenges in the quest for keystone. Bioscience 46:609–620
Rezende EL, Albert EM, Fortuna MA, Bascomte J (2009) Compartments in a marine food web associated with phylogeny, body mass, and habitat structure. Ecol Lett 12:779–788
Roberts A, Tregonning K (1980) The robustness of natural systems. Nature 288:265–266
Schoener TW (1993) On the relative importance of direct versus indirect effects in ecological communities. In: Kawanabe H, Cohen JE, Iwasaki K (eds) Mutualism and community organization: behavioral, theoretical, and food-web approaches. Oxford University Press, New York, pp 365–411
Stouffer DB, Bascompte J (2011) Compartmentalization increases food-web persistence. Proc Natl Acad Sci U S A 108:3648–3652
Strong DR (1992) Are trophic cascades all wet? Differentiation and donor-control in speciose ecosystems. Ecology 73:747–754
Teng J, McCann KS (2004) Dynamics of compartmented and reticulate food webs in relation to energetic flows. Am Nat 164:85–100
Tilman D, Kareiva P (1997) Spatial ecology: the role of space in population dynamics and interspecific interactions. Princeton University Press, Princeton
Winemiller KO (1990) Spatial and temporal variation in tropical fish trophic networks. Ecol Monogr 60:331–367
Yodzis P (1982) The compartmentation of real and assembled ecosystems. Am Nat 120:551–570
Yodzis P (1988) The indeterminancy of ecological interactions as perceived through perturbation experiments. Ecology 69:508–515

Part II
Inter-Specific Trait and Functional Diversity

Chapter 5
Direct and Indirect Effects of Facilitation and Competition in Ecological Communities: Population Persistence, Coexistence, and Species Diversity Along the Stress Gradients

Gaku Takimoto

Abstract The stress gradient hypothesis (SGH) predicts that pairwise species interactions within a single trophic level transition from negative to positive as the relative importance of facilitation to competition increases along the gradient of environmental severity. The SGH has inspired much empirical research testing the prediction, and the results have been utilized to develop more refined predictions. I develop a theoretical framework using mathematical models to incorporate these refined SGH predictions and investigate their population- and community-level consequences on how competition-facilitation transitions along the stress gradient affect population persistence, species coexistence, and species diversity in ecological communities. The framework suggests that population size and persistence of benefactor species as well as the sign and strength of interactions are important prerequisite for their facilitative and competitive interactions to have impacts on the population size, dynamics, and persistence of beneficiary species. At community level, in addition to the direct effect of benefactor species, indirect facilitation and indirect competition can play important roles in determining the coexistence of beneficiary species. The framework predicts that indirect competition tends to act against species coexistence and reduce species diversity in stressful environments, whereas indirect facilitation tends to promote species coexistence and foster high species diversity in benign environments. Moreover, it is shown that the indirect effects of benefactor species can be decomposed into density- and trait-mediated indirect effects, in a similar way to indirect effects mediated by trophic interactions. When competition-facilitation transition follows the classical SGH prediction, benefactor species generally promote the coexistence and diversity of beneficiary species both in stressful and in benign environments. A refined SGH prediction suggests stronger facilitation for stress-intolerant superior competitors. Incorporating such individual-level interactions tends to cause coexistence breakdown and the

G. Takimoto (✉)
Graduate School of Agricultural and Life Sciences, The University of Tokyo, Tokyo, Japan
e-mail: gakut@es.a.u-tokyo.ac.jp

A. Mougi (ed.), *Diversity of Functional Traits and Interactions*, Theoretical Biology,
https://doi.org/10.1007/978-981-15-7953-0_5

reduction of species diversity. Another refined SGH prediction is that benefactors exert competitive, rather than facilitative, effect in severest environments. Assuming such interactions results in that the range of stress factors permitting coexistence shifts toward more benign environmental conditions. Testing these population- and community-level predictions in the field may be useful to further our understanding on how facilitative species interactions contribute to the organization of ecological communities.

Keywords Competition · Facilitation · Stress gradient hypothesis · Indirect competition · Indirect facilitation

5.1 Introduction

Abiotic environments and biological interactions are important factors structuring ecological communities. Traditionally, interspecific competition has been considered as important biological interactions organizing communities of species belonging to the same trophic guilds. In contrast, facilitation—interspecific positive interactions between members of the same guilds—has received less attention. However, research in the past decades has accumulated evidence that facilitation can play major roles in the development and maintenance of ecological communities (Bruno et al. 2003; Brooker et al. 2008; Callaway 2007). In particular, the stress gradient hypothesis (SGH) has made a large contribution by motivating empirical research to test its prediction (Bertness and Callaway 1994). The prediction states that facilitation occurs more frequently in ecological communities in more stressful environments. This is a straightforward expectation because facilitative effects of benefactor species, such as amelioration of environmental conditions and provisioning of scarce resources, should be more effective and important than their competitive effects in more stressful, resource-lean environments.

Field observation and experiments testing the SGH prediction, however, found mixed results (Callaway 2007). These equivocal results are due in part to different types of stress factors (Kawai and Tokeshi 2007; Michalet 2007; Saccone et al. 2009), life history characteristics of benefactor and beneficiary species (Michalet 2007; Maestre et al. 2009), an incomplete coverage of the whole stress gradients, and others. These mixed findings motivated refinements of the SGH. Michalet et al. (2006) argued that facilitation becomes weak in extremely severe environments. They also emphasized that the stress levels at which transition from competition to facilitation occurs along the stress gradients depend on life history characteristics of beneficiary species (Choler et al. 2001; Liancourt et al. 2005; Michalet et al. 2006). Moreover, Michalet (2007) suggested that the interaction between types of stress factors (i.e., resource stress and non-resource stress) and life history characteristics of benefactor-beneficiary species pairs is important to determine strength of competition and facilitation along the stress gradients. Along this line, Maestre et al. (2009) generated a series of predictions regarding how types of stress factors and life history characteristics of benefactor-beneficiary species pairs interact to affect competition-facilitation transitions along the stress gradients (table 1 in Maestre et al. 2009). In particular, their predictions considered that the effect of benefactor species can become negative under severe environmental conditions (in contrast to positive, as predicted by the original SGH) when severe resource competition between benefactors and beneficiaries occurs under extremely resource-deficient conditions. A graphical model by Holmgren and Scheffer (2010) illustrated a potential mechanism of facilitation maximized at moderate stress levels. More recent empirical works have found supportive evidence for these refined SGH predictions (Graff and Aguiar 2017; O'Brien et al. 2017; Scrosati 2017; Qi et al. 2018; Zhang et al. 2018).

On a community scale, transition between negative and positive interactions can influence community organization along the stress gradients. The early model by Grime (1973), with no regard to facilitation but assuming a dominant role of competition, predicted that species diversity of plant communities is heightened at intermediate stress levels, stress-intolerant superior competitors dominate and lower species diversity at lower stress levels, and only stress-tolerant inferior competitors can persist in low diversity communities under severe environments. Hacker and Gaines (1997) incorporated facilitation into the Grime's model. Their model suggested that facilitation, expectedly stronger in severer environments, should enhance species diversity at intermediate to high stress levels. Michalet et al. (2006) revised the Hacker and Gaines' model in that facilitation may be diminished and thus unable to elevate species diversity in extremely severe environments. These pioneer models (Grime 1973; Hacker and Gaines 1997; Michalet et al. 2006) have used graphical approaches. In contrast, Xiao et al. (2009) and Le Bagousse-Pinguet et al. (2014) employed stochastic simulation models to test mathematically predictions of the earlier graphical models. These simulation models, assuming that facilitation became stronger with environmental severity but vanished in extremely severe environments, confirmed the prediction of Michalet et al. (2006).

In this chapter, I aim to develop a theoretical model that incorporates recent advanced understandings on the relationships between stress levels and competition-facilitation transitions to generate community-level predictions on species coexistence and diversity. As in Xiao et al. (2009) and Le Bagousse-Pinguet et al. (2014), I use a mathematical approach. Although their simulation models have made simplifying assumptions that (1) facilitative effects do not depend on the identity of benefactor and beneficiary species, and (2) competitive effects do not vary with environmental severity, the current framework relaxes these assumptions to consider a wider range of refined SGH predictions regarding competition-facilitation transitions along the gradient of environmental severity. In addition, instead of using stochastic simulations, I formulate the current model using a set of differential equations describing the population dynamics of benefactors and beneficiaries. This approach is less computationally challenging, and suited to the present purpose of comparing how several different modes of individual-level interactions scale up to change population- and community-level patterns of species coexistence and species diversity.

5.2 Framework Development

5.2.1 The Patch Occupancy Model

The model assumes that a tradeoff between competitive and stress-tolerant abilities constrains life history strategies of beneficiary species j ($j = 1, \cdots, M$). There is transitive competitive relationships among beneficiary species; a species with a smaller species index j is a superior competitor to a species with a larger index j' ($j < j'$). On the other hand, species with a larger index is more stress tolerant than species with a smaller index. Species with higher stress tolerance can persist under higher stress levels without facilitative or competitive effects from other species. I use the standard patch occupancy model (Hastings 1980; Tilman 1994) as a basis to build the model for the population dynamics of beneficiary species. I assume that a single individual of beneficiary species can occupy a patch (another individual of benefactor species can share the same patch; see below). In this model, when an individual of higher ranked species (with smaller species index) colonizes into a patch occupied by an individual of lower ranked species, the individual of higher ranked species excludes and replaces the individual of the lower ranked (competitive relationships among beneficiary species are transitive).

Let us denote the frequency of patches occupied by a beneficiary species j as y_j. Then the model is described as:

$$\frac{dy_j}{dt} = u_j^* y_j \left(1 - \sum_{k=1}^{j} y_k\right) - v_j^* y_j - \left(\sum_{k=1}^{j-1} u_k^* y_k\right) y_j \tag{5.1}$$

where u_j^* and v_j^* are the colonization and mortality rates of the species j, respectively. To incorporate the effect of benefactor on beneficiary species, I consider that benefactor species modify the colonization and mortality rates of beneficiary species when they occur in the same patch.

$$u_j^* = u_j \lambda \left(1 - \sum_{i=1}^{N} R_i\right) + \sum_{i=1}^{N} \left(u_j \lambda \frac{1 + F_{ji}^u}{1 + C_{ji}^u} R_i\right) \tag{5.2a}$$

$$v_j^* = v_i \sigma \left(1 - \sum_{i=1}^{N} R_i\right) + \sum_{i=1}^{N} \left(v_j \sigma \frac{1 + C_{ji}^v}{1 + F_{ji}^v} R_i\right) \tag{5.2b}$$

where R_i denotes the frequency of patches occupied by the benefactor species i ($i = 1, \cdots, N$). The colonization rate is composed of two terms: $u_j \lambda \left(1 - \sum_{i=1}^{N} R_i\right)$ and $\sum_{i=1}^{N} \left(u_j \lambda \frac{1+F_{ji}^u}{1+C_{ji}^u} R_i\right)$. The first term represents the colonization rate from a patch occupied only by the beneficiary species j (no co-occurrence of benefactor species).

Colonization from such a patch occurs with the baseline colonization rate $u_j\lambda$, where u_j is the species-specific colonization coefficient and λ is a stress factor affecting the colonization rate. This baseline colonization rate is multiplied by the probability that a patch has no benefactor species is $1 - \sum_{i=1}^{N} R_i$ to define the first term. The second term is the sum of colonization rates from a patch where the beneficiary species j co-occurs with a benefactor species. When a beneficiary species j occurs with the benefactor species i, its colonization rate is increased or decreased from the baseline rate $u_j\lambda$ due to species i's facilitative effect F_{ji}^u or competitive effect C_{ji}^u, respectively, on species j, which is multiplied by the probability R_i that the patch has a co-occurring benefactor species i.

The mortality rates are similarly composed of two terms; $v_i\sigma\left(1 - \sum_{i=1}^{N} R_i\right)$ is the mortality rate of individuals occupying a patch by itself (without benefactor species), and $\sum_{i=1}^{N}\left(v_j\sigma\frac{1+C_{ji}^v}{1+F_{ji}^v}R_i\right)$ is the sum of mortality rates of individuals sharing patches with benefactor species. The baseline mortality rate is $v_j\sigma$, where v_j is the species-specific mortality coefficient and σ is the another stress factor determining the mortality. In the presence of benefactor species, the mortality rate is increased due to benefactor's competitive effect C_{ji}^v, and decreased due to its facilitative effect F_{ji}^v.

Thus, this model considers two types of stress factors: λ and σ. The smaller the colonization stress factor λ, the higher the colonization stress and the lower the baseline colonization rate $u_j\lambda$. On the other hand, the larger the mortality stress factor σ, the higher the mortality stress and the higher the baseline mortality rate $v_j\sigma$.

The quotient $\frac{\sigma}{\lambda}$ measures the severity of combined colonization and mortality stress; larger values of $\frac{\sigma}{\lambda}$ mean severer combined stress. Colonization and mortality coefficients, u_i and v_i, respectively, are chosen to reflect the competition-stress-tolerance tradeoff; species of a larger index are competitively inferior but more stress tolerant, and can persist under severer combined stress (i.e., higher values of $\frac{\sigma}{\lambda}$) when there is no interspecific effects from other species: $\frac{u_1}{v_1} < \frac{u_2}{v_2} < \cdots < \frac{u_M}{v_M}$.

I define patch frequency dynamics for benefactor species in a similar manner:

$$\frac{dR_i}{dt} = (U_i\lambda)R_i\left(1 - \sum_{k=1}^{i} R_k\right) - (V_i\sigma)R_i - \left(\sum_{k=1}^{i-1}(U_k\lambda)R_k\right)R_i \tag{5.3}$$

As with beneficiary species, a tradeoff between competitive ability and stress tolerance constrains the life history parameters of benefactor species. The same colonization and mortality stress factors σ and λ affect the colonization and mortality rates, with species-specific colonization and morality coefficients: U_i and V_i, respectively. These coefficients are chosen to reflect the tradeoff: $\frac{U_1}{V_1} < \frac{U_2}{V_2} < \cdots < \frac{U_N}{V_N}$. There is no influence of beneficiary species on benefactor species.

5.3 Analysis and Results

In this section, I first illustrate how different competition-facilitation transitions along the environmental severity gradients can be implemented in the above model (*Competition and Facilitation between Individuals*). Next, to explore population-level expressions of competition-facilitation transitions along the stress gradients, I use a simplified version of the model for a one-benefactor-one-beneficiary system and show that benefactor's competitive and facilitative effects on beneficiaries depend on benefactor's population abundance along the stress gradients (*Effect of Benefactor Abundance on Population Persistence of Beneficiaries*). I then turn to a one-benefactor-two-beneficiary system to explore how benefactor's impacts propagate through multiple beneficiary species, and illustrate that benefactor's net effect on a beneficiary species can be decomposed into direct effect and indirect effect mediated by another beneficiary species (*Indirect Effects of Benefactor Species on Beneficiary Species*). These direct and indirect effects are shown to drive benefactor's impacts on the coexistence of beneficiary species (*Indirect Facilitation, Indirect Competition, and the Coexistence of Beneficiary Species*). Finally, using a one-benefactor-many-beneficiary system, I explore that the diversity of beneficiary species responds to environmental severity gradients, and explain that the mechanisms by which a benefactor promotes/inhibits the coexistence of beneficiaries drive these diversity patterns under different modes of competition-facilitation transitions along the stress gradients (*Species Diversity along the Stress Gradients*).

Model variables and important parameters used in the above model and following analyses are explained in Tables 5.1 and 5.2. Parameter values used in the analyses are noted in figure legends. I confirmed that qualitative conclusions do not hinge on specific parameter values used in presented figures.

5.3.1 Competition and Facilitation Between Individuals

Effects of interactions between individuals occupying the same patch are expressed by the terms, F^u_{ji}, C^u_{ji}, F^v_{ji}, and C^v_{ji}, in Eq. (5.2b). Strength of these facilitative and competitive effects may vary along the stress gradients. Thus, these terms may be the functions of the stress factors λ and σ:

$$F^u_{ji} = F^u_{ji}(\lambda, \sigma) \tag{5.4a}$$

$$C^u_{ji} = C^u_{ji}(\lambda, \sigma) \tag{5.4b}$$

for the colonization rates, and

Table 5.1 List of model variables and parameters

Symbol	Definition
y_j	Patch frequency of *j*th beneficiary species
R_i	Patch frequency of *i*th benefactor species
M	Number of beneficiary species
N	Number of benefactor species
u_j^*	Colonization rate of beneficiary species *j*
λ	Colonization stress
σ	Mortality stress
v_j^*	Mortality rate of beneficiary species *j*
u_j	Baseline colonization coefficient of beneficiary species *j* (without benefactor's effect)
v_j	Baseline mortality coefficient of beneficiary species *j* (without benefactor's effect)
U_i	Colonization coefficient of benefactor species *i*
V_i	Mortality coefficient of benefactor species *i*
F_{ji}^u	Benefactor *i*'s facilitative effect on beneficiary *j*'s colonization
C_{ji}^u	Benefactor *i*'s competitive effect on beneficiary *j*'s colonization
F_{ji}^v	Benefactor *i*'s facilitative effect on beneficiary *j*'s mortality
C_{ji}^v	Benefactor *i*'s competitive effect on beneficiary *j*'s mortality
k^u	Parameter to control the strength of benefactor's effects on beneficiary's colonization; larger values represent larger facilitative and smaller competitive effects
k^v	Parameter to control the strength of benefactor's effects on beneficiary's mortality; larger values represent larger facilitative and smaller competitive effects
f_{ji}^{uX}	Benefactor *i*'s facilitative effect on beneficiary *j*'s colonization in extremely stressful environments (for cases IIA and B)
c_{ji}^{uX}	Benefactor *i*'s competitive effect on beneficiary *j*'s colonization in extremely stressful environments (for cases IIA and B)
f_{ji}^{vX}	Benefactor *i*'s facilitative effect on beneficiary *j*'s mortality in extremely stressful environments (for cases IIA and B)
c_{ji}^{vX}	Benefactor *i*'s competitive effect on beneficiary *j*'s mortality in extremely stressful environments (for cases IIA and B)
X_λ	Extreme colonization stress level around which benefactor's facilitation and competitive effects shift to f_{ji}^{uX} and c_{ji}^{uX}, respectively (for cases IIA and B)
X_σ	Extreme mortality stress level around which benefactor's facilitation and competitive effects shift to f_{ji}^{vX} and c_{ji}^{vX}, respectively (for cases IIA and B)
q_u	Rate at which benefactor's facilitation and competitive effects shift to f_{ji}^{uX} and c_{ji}^{uX}, respectively, in extremely stressful environments (for cases IIA and B)
q_v	Rate at which benefactor's facilitation and competitive effects shift to f_{ji}^{vX} and c_{ji}^{vX}, respectively, in extremely stressful environments (for cases IIA and B)

$$F_{ji}^v = F_{ji}^v(\lambda, \sigma) \tag{5.4c}$$

$$C_{ji}^v = C_{ji}^v(\lambda, \sigma) \tag{5.4d}$$

for the mortality rates. These functions can take various forms to incorporate different kinds of transition between facilitation and competition along the stress gradient.

Table 5.2 The facilitative and competitive effects of benefactor species *i* on the colonization and mortality of beneficiary species *j* in four representative cases of competition-facilitation transitions along the stress gradients

	F_{ji}^{u}	C_{ji}^{u}	F_{ji}^{v}	C_{ji}^{v}
Case IA	$\frac{f_{ji}^{u}}{\lambda}\left(=\frac{k^{u}}{\lambda}\right)$	$c_{ji}^{u}\lambda\left(=\frac{\lambda}{k^{u}}\right)$	$f_{ji}^{v}\sigma\left(=\frac{\sigma}{k^{v}}\right)$	$\frac{c_{ji}^{v}}{\sigma}\left(=\frac{k^{v}}{\sigma}\right)$
Case IB	$\frac{f_{ji}^{u}}{\lambda}\left(=\frac{k^{u}}{u_j\lambda}\right)$	$c_{ji}^{u}\lambda\left(=\frac{u_j\lambda}{k^{u}}\right)$	$f_{ji}^{v}\sigma\left(=\frac{v_j\sigma}{k^{v}}\right)$	$\frac{c_{ji}^{v}}{\sigma}\left(=\frac{k^{v}}{v_j\sigma}\right)$
Case IIA	$\left(f_{ji}^{u}/\lambda\right)\Omega_\lambda+f_{ji}^{uX}(1-\Omega_\lambda)$ $\left(=\left(\frac{k^{u}}{\lambda}\right)\Omega_\lambda+f_{ji}^{uX}(1-\Omega_\lambda)\right)$	$\left(c_{ji}^{u}\lambda\right)\Omega_\lambda+c_{ji}^{uX}(1-\Omega_\lambda)$ $\left(=\left(\frac{\lambda}{k^{u}}\right)\Omega_\lambda+c_{ji}^{uX}(1-\Omega_\lambda)\right)$	$\left(f_{ji}^{v}\sigma\right)(1-\Omega_\sigma)+f_{ji}^{vX}\Omega_\sigma$ $\left(=\left(\frac{\sigma}{k^{v}}\right)(1-\Omega_\sigma)+f_{ji}^{vX}\Omega_\sigma\right)$	$\left(c_{ji}^{v}/\sigma\right)(1-\Omega_\sigma)+c_{ji}^{vX}\Omega_\sigma$ $\left(=\left(\frac{k^{v}}{\sigma}\right)(1-\Omega_\sigma)+c_{ji}^{vX}\Omega_\sigma\right)$
Case IIB	$\left(f_{ji}^{u}/\lambda\right)\Omega_\lambda+f_{ji}^{uX}(1-\Omega_\lambda)$ $\left(=\left(\frac{k^{u}}{u_j\lambda}\right)\Omega_\lambda+f_{ji}^{uX}(1-\Omega_\lambda)\right)$	$\left(c_{ji}^{u}\lambda\right)\Omega_\lambda+c_{ji}^{uX}(1-\Omega_\lambda)$ $\left(=\left(\frac{u_j\lambda}{k^{u}}\right)\Omega_\lambda+c_{ji}^{uX}(1-\Omega_\lambda)\right)$	$\left(f_{ji}^{v}\sigma\right)(1-\Omega_\sigma)+f_{ji}^{vX}\Omega_\sigma$ $\left(=\left(\frac{v_j\sigma}{k^{v}}\right)(1-\Omega_\sigma)+f_{ji}^{vX}\Omega_\sigma\right)$	$\left(c_{ji}^{v}/\sigma\right)(1-\Omega_\sigma)+c_{ji}^{vX}\Omega_\sigma$ $\left(=\left(\frac{k^{v}}{v_j\sigma}\right)(1-\Omega_\sigma)+c_{ji}^{vX}\Omega_\sigma\right)$

Note: $\Omega_\lambda=\frac{\lambda^{q_u}}{\lambda^{q_u}+X_\lambda^{q_u}}$ and $\Omega_\sigma=\frac{\sigma^{q_v}}{\sigma^{q_v}+X_\sigma^{q_v}}$ in cases IIA and B determine the stress level at which the sign of interaction shifts from positive to negative or neutral in extremely stressful environments with very small λ and/or very large σ

Here I illustrate four representative cases, IA, IB, IIA, and IIB, as examples of competition-facilitation transitions, which will be used later for population- and community-level analysis. Case IA corresponds to the original SGH prediction, in which competition predominates in low stress environments, and facilitation is more frequent in severer environments (Bertness and Callaway 1994). Possible functional forms for such a situation are

$$F_{ji}^{u} = f_{ji}^{u}/\lambda \tag{5.5a}$$

$$C_{ji}^{u} = c_{ji}^{u}\lambda \tag{5.5b}$$

for the colonization rates, and

$$F_{ji}^{v} = f_{ji}^{v}\sigma \tag{5.5c}$$

$$C_{ji}^{v} = c_{ji}^{v}/\sigma \tag{5.5d}$$

for the mortality rates. For the colonization rates, the increase of stress factor λ (i.e., weaker stress) decreases the facilitation effect F_{ji}^{u} and increases the competition effect C_{ji}^{u}. For the mortality rates, the increase of the stress factor σ (severer stress) increases the facilitation effect F_{ji}^{v} and decreases the competition effect C_{ji}^{v}. The relative strength of these facilitative and competitive effects determines the net effect of a co-occurring benefactor species *i*. Specifically, for example, the colonization and mortality rates in the presence of benefactor species stays constant across the entire stress gradient if $f_{ji}^{u}c_{ji}^{u} = 1$ and $f_{ji}^{v}c_{ji}^{v} = 1$, respectively. I use a specific case of this, in which I set the effect parameters as follows:

$$f_{ji}^{u} = 1/c_{ji}^{u} = k^{u} \tag{5.6a}$$

$$1/f_{ji}^{v} = c_{ji}^{v} = k^{v} \tag{5.6b}$$

Under this formulation, stress-tolerant inferior competitor species receive stronger positive effects from the benefactor, and transition of the colonization and mortality rates from competition to facilitation with intensifying stress occurs at the same stress levels for all beneficiary species (Fig. 5.1a, b).

A number of empirical evidence suggest that life history strategy of beneficiary species can influence the stress levels at which benefactor's net effect transitions from negative to positive (Choler et al. 2001; Liancourt et al. 2005; Michalet et al. 2006). Specifically, stress-intolerant superior competitors tend to receive facilitative effects in more benign environments as compared to stress-tolerant inferior competitors. Case IB incorporates this by the following formulation:

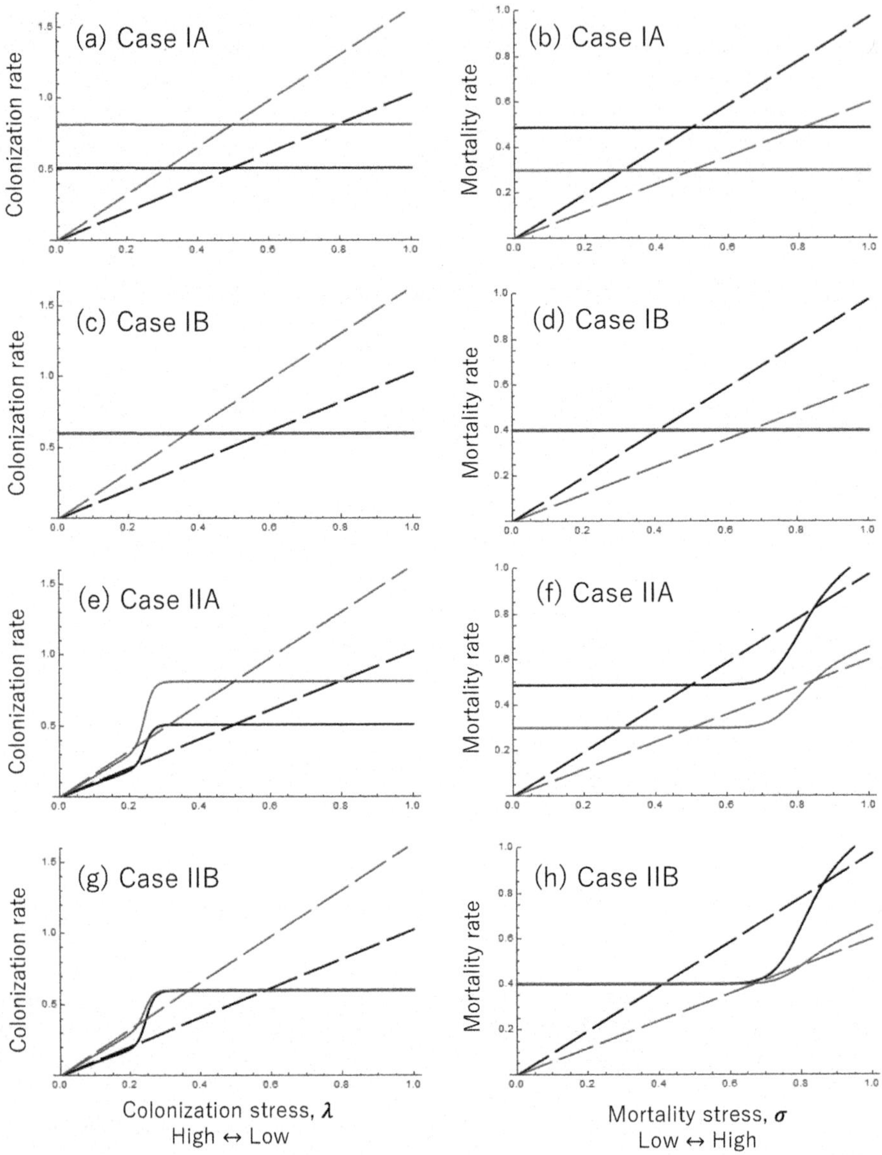

Fig. 5.1 Transition of benefactor's direct effect between facilitation and competition on the colonization and mortality rates of beneficiary species. Lines represent colonization rates (u_j^*) or mortality rates (v_j^*) in patches without benefactor species ($\sum_{i=1}^{N} R_i = 0$, dashed lines), or those with benefactor species ($\sum_{i=1}^{N} R_i = 1$, solid lines). Black lines are for a stress-intolerant superior competitor ($j = 1$), and gray for a stress-tolerant inferior competitor ($j = 20$). Parameter values are $k^u = 0.5$ for (**a**) and (**e**), and 0.6 for (**c**) and (**g**); $k^v = 0.5$ for (**b**) and (**f**), and 0.4 for (**d**) and (**h**). Other parameter

$$f_{ji}^{u} = 1/c_{ji}^{u} = k^{u}/u_{j} \tag{5.7a}$$

$$1/f_{ji}^{v} = c_{ji}^{v} = k^{v}/v_{j} \tag{5.7b}$$

This formulation sets the colonization and facilitation rates at the same values across all beneficiary species, which results in stress-tolerant inferior competitors receive net positive effects from benefactors at severer stress levels than stress-intolerant superior competitors (Fig. 5.1c, d).

More recent refinement of the hypothesis suggests that intensifying stress causes transition from competition to facilitation, but facilitation is weakened and net effect from benefactors eventually vanishes (Michalet et al. 2006) or becomes negative as a result of competition for scarce resource (Tielbörger and Kadmon 2000; Maestre and Cortina 2004) in severest environments. Such transition may be formulated as follows:

$$F_{ji}^{u} = \left(f_{ji}^{u}/\lambda\right)\Omega_{\lambda} + f_{ji}^{uX}(1 - \Omega_{\lambda}) \tag{5.8a}$$

$$C_{ji}^{u} = \left(c_{ji}^{u}\lambda\right)\Omega_{\lambda} + c_{ji}^{uX}(1 - \Omega_{\lambda}) \tag{5.8b}$$

for the colonization rates, and

$$F_{ji}^{v} = \left(f_{ji}^{v}\sigma\right)(1 - \Omega_{\sigma}) + f_{ji}^{vX}\Omega_{\sigma} \tag{5.8c}$$

$$C_{ji}^{v} = \left(c_{ji}^{v}/\sigma\right)(1 - \Omega_{\sigma}) + c_{ji}^{vX}\Omega_{\sigma} \tag{5.8d}$$

for the mortality rates. The terms Ω_{λ} and Ω_{σ} control the stress level at which the sign of interaction shifts from positive to negative (e.g., $f_{ji}^{uX} < c_{ji}^{uX}$ and $f_{ji}^{vX} < c_{ji}^{vX}$) or to neutral ($f_{ji}^{uX} = c_{ji}^{uX}$ and $f_{ji}^{vX} = c_{ji}^{vX}$) in extremely severe environments (very small λ and very large σ):

$$\Omega_{\lambda} = \frac{\lambda^{q_u}}{\lambda^{q_u} + X_{\lambda}^{q_u}} \tag{5.9a}$$

$$\Omega_{\sigma} = \frac{\sigma^{q_v}}{\sigma^{q_v} + X_{\sigma}^{q_v}} \tag{5.9b}$$

where $X_{\lambda}^{q_u}$ and $X_{\sigma}^{q_v}$ determine the stress levels at which these transitions occur and q_u and q_v control how quickly the transitions occur. In parallel with cases IA and IB, I consider two cases IIA (Fig. 5.1e, f) and IIB (Fig. 5.1g, h) as differing in terms of

Fig. 5.1 (continued) values are common for all relevant cases: $u_j = (1.05)^{\frac{j}{2}}$, $v_j = (0.95)^{\frac{j}{2}}$, $f_{ji}^{uX} = 0$, $c_{ji}^{uX} = 0.1$, $f_{ji}^{vX} = 0$, $c_{ji}^{vX} = 0.1$, $X_{\lambda} = 0.25$, $X_{\sigma} = 0.75$, $q_u = 20$, and $q_v = 20$

how the stress levels at which the net effect of benefactors changes from competition to facilitation depend on the life history strategy of beneficiary species.

5.3.2 *Effect of Benefactor Abundance on Population Persistence of Beneficiaries*

The net effect of a benefactor species on the population of a beneficiary species depends not only on the strength of individual interactions per se but also on the abundance of the benefactor species. This is important because, in addition to the strength of facilitative and competitive interactions at the individual level, the population abundance of benefactor species may change across the stress gradient, affecting the sign and strength of net effects at the population level.

To demonstrate this, we rewrite the model (1) and (3) for population dynamics of one benefactor species and one beneficiary species:

$$\frac{dR}{dt} = U\lambda R(1 - R) - V\sigma R \tag{5.10a}$$

$$\frac{dy}{dt} = u\lambda\left(1 + \left(\frac{F^u - C^u}{1 + C^u}\right)R\right)y(1 - y) - v\sigma\left(1 + \left(\frac{C^v - F^v}{1 + F^v}\right)R\right)y \tag{5.10b}$$

where we drop the subscripts of species indices for simplicity. We focus on the equilibrium abundances:

$$\overline{R} = \begin{cases} 0 & \left(\text{when } \frac{\sigma}{\lambda} \geq \frac{U}{V}\right) \\ 1 - \frac{V}{U}\cdot\frac{\sigma}{\lambda} & \left(\text{when } \frac{\sigma}{\lambda} < \frac{U}{V}\right) \end{cases} \tag{5.11a}$$

and

$$\overline{y} = \begin{cases} 0 & \left(\text{when } \frac{\sigma}{\lambda} \geq \frac{u}{v}\cdot\frac{1 + \left(\frac{F - C^u}{1 + C^u}\right)\overline{R}}{1 + \left(\frac{C^v - F^v}{1 + F^v}\right)\overline{R}}\right) \\ 1 - \frac{v}{u}\cdot\frac{1 + \left(\frac{C - F^v}{1 + F^v}\right)\overline{R}}{1 + \left(\frac{F^u - C^u}{1 + C^u}\right)\overline{R}}\cdot\frac{\sigma}{\lambda} & \left(\text{when } \frac{\sigma}{\lambda} < \frac{u}{v}\cdot\frac{1 + \left(\frac{F - C^u}{1 + C^u}\right)\overline{R}}{1 + \left(\frac{C^v - F^v}{1 + F^v}\right)\overline{R}}\right) \end{cases} \tag{5.11b}$$

These equilibrium abundances indicate that stress levels for the persistence of the beneficiary species depends both on the facilitative and competitive effects of the benefactor species as well as its abundance, which itself depends on stress levels.

This is visualized by plotting the abundance of the beneficiary species together with the abundance of the benefactor species (Fig. 5.2). If the benefactor species are relatively stress tolerant and can persist across a wide stress gradient, its net effect can affect the persistence as well as abundance of the beneficiary species (Fig. 5.2a, b for cases IA and IB, and Fig. 5.2e, f for cases IIA and IIB). On the other hand, if the benefactor species are not very stress tolerant and can occur only within a limited range of the stress gradient, its net effect on the beneficiary species is also limited within this range (Fig. 5.2c, d for cases IA and IB, and Fig. 5.2g, h for cases IIA and IIB).

5.3.3 *Indirect Effects of Benefactor Species on Beneficiary Species*

Facilitative and competitive effects from a single benefactor species can influence multiple beneficiary species. When these beneficiary species interact each other, they can transmit indirect effects of benefactor species to other beneficiary species (Levine 1999; Brooker et al. 2008; Xiao and Michalet 2013). For example, when a benefactor has a direct facilitative effect on a beneficiary species, this benefactor may facilitate another beneficiary species at the same time, which can in turn lower the abundance of the first beneficiary species through competitive interaction (Maestre et al. 2004; Michalet et al. 2015).

To analyze the indirect effect of benefactor species on beneficiary species. I use a simplified version of the model (1) and (3) to describe the population dynamics of one benefactor and two beneficiary species.

$$\frac{dR}{dt} = U\lambda R(1 - R) - V\sigma R \tag{5.12a}$$

$$\frac{dy_1}{dt} = u_1\lambda\left(1 + \left(\frac{F_1^u - C_1^u}{1 + C_1^u}\right)R\right)y_1(1 - y_1) - v_1\sigma\left(1 + \left(\frac{C_1^v - F_1^v}{1 + F_1^v}\right)R\right)y_1 \tag{5.12b}$$

$$\begin{aligned}\frac{dy_2}{dt} &= u_2\lambda\left(1 + \left(\frac{F_2^u - C_2^u}{1 + C_2^u}\right)R\right)y_2(1 - y_1 - y_2) \\ &\quad - u_1\lambda\left(1 + \left(\frac{F_1^u - C_1^u}{1 + C_1^u}\right)R\right)y_1y_2 - v_2\sigma\left(1 + \left(\frac{C_2^v - F_2^v}{1 + F_2^v}\right)R\right)y_2\end{aligned} \tag{5.12c}$$

In this model, the indirect effect of a benefactor species is mediated by the beneficiary species 1 and transmitted to the beneficiary species 2. The beneficiary species 1 is a stress-intolerant superior competitor, and the beneficiary species 2 is a stress-tolerant inferior competitor.

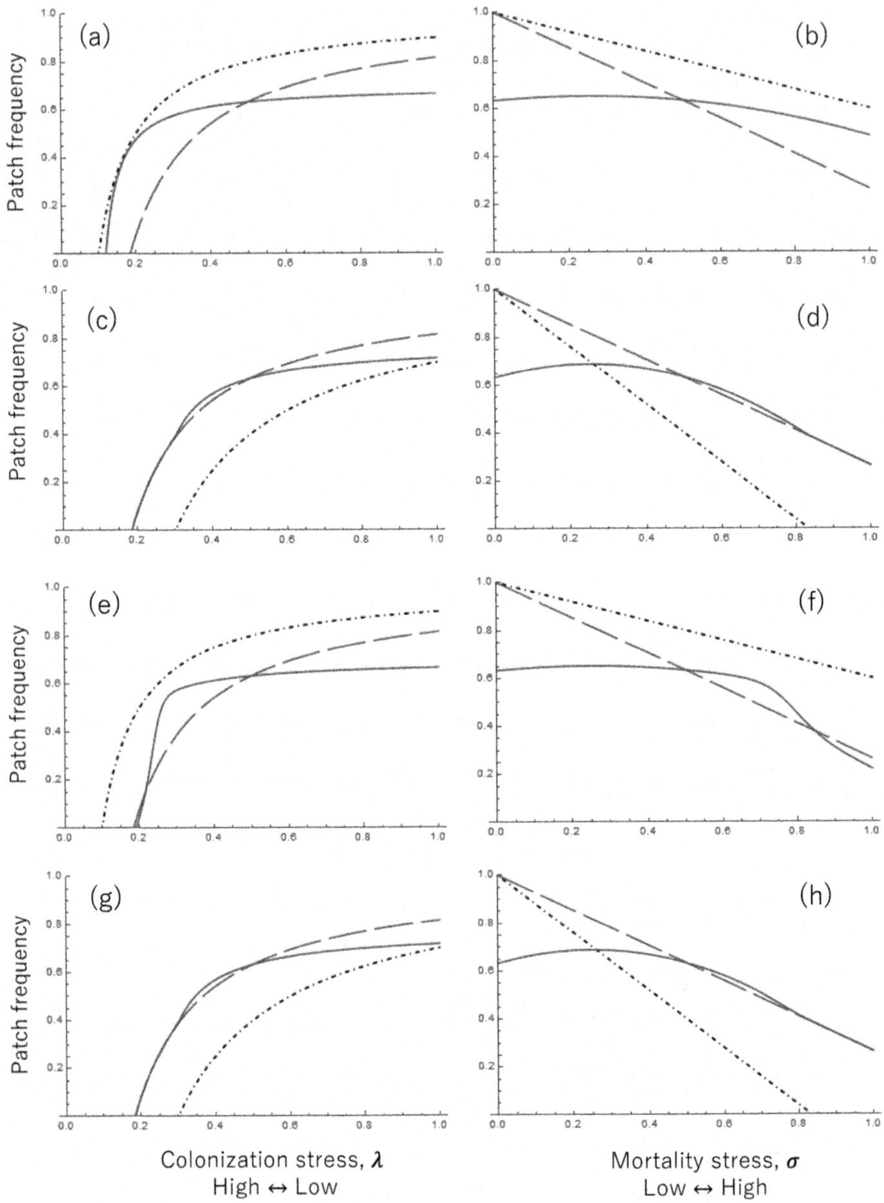

Fig. 5.2 Effect of benefactor's abundance on the abundance of a beneficiary species. Dot-dashed lines are the patch frequency of the benefactor species. Dashed lines are the patch frequency of a beneficiary species ($j = 20$) without the benefactor species, and solid lines are with the benefactor species. Parameter values are $V = 0.2$ for (**a**), (**b**), (**e**), and (**f**), and 0.6 for (**c**), (**d**), (**g**), and (**h**); $U = 1$, $k^{u} = 0.5$, $k^{v} = 0.5$ for all panels (**a–h**). Other parameter values are the same as in Fig. 5.1

To evaluate the effects of the benefactor species, I compare the equilibrium abundances of beneficiary species between with and without the benefactor. Denote the equilibrium abundances of the beneficiary species 1 with and without the benefactor species as $\overline{y}_1$ and $\overline{y}_{1,R=0}$, respectively. Denote those of the beneficiary species 2 with and without the benefactor species in the absence of the beneficiary species 1 as $\overline{y}_{2,y_1=0}$ and $\overline{y}_{2,y_1=0,R=0}$, respectively. Denote those of the beneficiary species 2 in the presence of the beneficiary species 1 with and without the benefactor species as $\overline{y}_2$ and $\overline{y}_{2,R=0}$, respectively. The differences in equilibrium abundances between with and without the benefactors can be calculated as follows:

$$\overline{y}_1 - \overline{y}_{1,R=0} = \frac{v_1}{u_1}\frac{\sigma}{\lambda}(1 - W_1) \tag{5.13a}$$

$$\overline{y}_{2,y_1=0} - \overline{y}_{2,y_1=0,R=0} = \frac{v_2}{u_2}\frac{\sigma}{\lambda}(1 - W_2) \tag{5.13b}$$

$$\overline{y}_2 - \overline{y}_{2,R=0} = \frac{v_2}{u_2}\frac{\sigma}{\lambda}(1 - W_2) - \left(1 + \frac{u_1}{u_2}X\right)\frac{v_1}{u_1}\frac{\sigma}{\lambda}(1 - W_1) + \frac{u_1}{u_2} \times (1 - X)\overline{y}_{1,R=0} \tag{5.13c}$$

where

$$W_1 = \frac{1 + \left(\frac{\frac{1}{C} - F_1^v}{1 + F_1^v}\right)\overline{R}}{1 + \left(\frac{F_1^u - C_1^u}{1 + C_1^u}\right)\overline{R}} \tag{5.14a}$$

$$W_2 = \frac{1 + \left(\frac{\frac{2}{C} - F_2^v}{1 + F_2^v}\right)\overline{R}}{1 + \left(\frac{F_2^u - C_2^u}{1 + C_2^u}\right)\overline{R}} \tag{5.14b}$$

$$X = \frac{1 + \left(\frac{\frac{1}{F} - C_1^u}{1 + C_1^u}\right)\overline{R}}{1 + \left(\frac{F_2^u - C_2^u}{1 + C_2^u}\right)\overline{R}} \tag{5.14c}$$

Our primary focus is on $\overline{y}_2 - \overline{y}_{2,R=0}$, the net effect of the benefactor species on the beneficiary species 2. The expression of this net effect has three terms. The first term $\frac{v_2}{u_2}\frac{\sigma}{\lambda}(1 - W_2)$ expresses the contribution of the direct effect of the benefactor on the beneficiary species 2 (cf. Eq. 5.13b). This direct effect is positive when $W_2 > 1$ and negative when $W_2 < 1$. The second term $-\left(1 + \frac{u_1}{u_2}X\right)\frac{v_1}{u_1}\frac{\sigma}{\lambda}(1 - W_1)$ is the indirect effect through the beneficiary species 1. Within this term, $\frac{v_1}{u_1}\frac{\sigma}{\lambda}(1 - W_1)$ expresses benefactor's direct effect on the beneficiary species 1, with the sign of $1 - W_1$ indicating the direction of the effect (cf. Eq. 5.13a). This direct effect is multiplied by $-\left(1 + \frac{u_1}{u_2}X\right)$, when transmitted as an indirect effect to the beneficiary species 2. The negative sign represents that the increased abundance of the beneficiary species 1 increases competitive suppression of the beneficiary species 2. This competitive

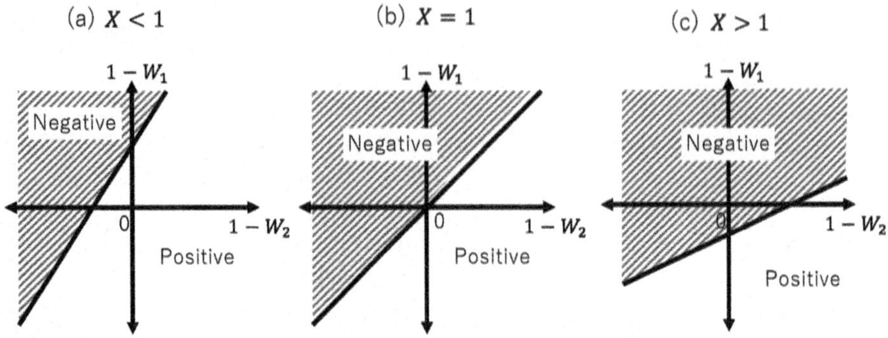

Fig. 5.3 Sign of indirect effects of a benefactor species on a beneficiary species. The transmitter is a stress-intolerant superior competitor (the beneficiary species 1), and the receiver of the indirect effects is a stress-tolerant inferior competitor (beneficiary species 2). The sign and magnitude of $1 - W_1$ indicate those of donor's (benefactor's) direct effects on the transmitter. The sign and magnitude of $1 - W_2$ indicate those of donor's direct effects on the receiver. X is the strength of donor's effect on transmitter's colonization rate relative to that of donor's effect on receiver's colonization rate. The donor-to-transmitter effect is weaker when $X < 1$ (**a**), is of the same strength when $X = 1$ (**b**), and is stronger when $X > 1$ (**c**), relative to the donor-to-receiver effect. The slope and y-intercept of the boundary between negative and positive indirect effects are $\frac{\frac{v_2}{u_2}}{\left(1+\frac{u_1}{u_2}X\right)\left(\frac{v_1}{u_1}\right)}$ and $\frac{\frac{u_1}{u_2}(1-X)\left(1-\frac{v_1}{u_1}\frac{q}{\lambda}\right)}{\left(1+\frac{u_1}{u_2}X\right)\left(\frac{v_1}{u_1}\right)\frac{q}{\lambda}}$, respectively

effect is modulated by the factor $1 + (u_1/u_2)X$, in which X means that benefactor's effects on beneficiary species' colonization rates control the competitive suppression. When $X > 1$, the benefactor causes a relative enhancement of the colonization rate of the beneficiary species 1 to that of the beneficiary species 2, which intensifies the competitive suppression. When $X < 1$, there is a relative decrease in the colonization rate of the beneficiary species 1 to that of the beneficiary species 2. This weakens the competitive suppression. The third term $\frac{u_1}{u_2}(1 - X)\bar{y}_{1,R=0}$ means that there is an indirect effect even when the benefactor does not change the abundance of the beneficiary species 1 (i.e., when $W_1 = 1$), as far as the benefactor modifies relative colonization rates of the beneficiary species (i.e., $X \neq 1$).

Interestingly, this interpretation of the net effect indicates that the beneficiary species 1 can mediate two types of indirect effects. One is a density-mediated indirect effect, caused by the benefactor changing the abundance of the species 1. The sign and magnitude of $1 - W_1$ determine the strength of this density-mediated indirect effect. The other is a trait-mediated indirect effect, caused by the benefactor species changing the colonization rate of the beneficiary species 1 relative to that of the beneficiary species 2. The value of X measures benefactor's effect on competitive advantage of the beneficiary species 1 over 2, and modulates this trait-mediated indirect effect on the beneficiary species 2. These density- and trait-mediated indirect effects are added to the direct effect on the beneficiary species 2 to determine the net effect. Figure 5.3 illustrates that a net negative effect is expected either when the direct negative effect on the beneficiary species 2 is large enough (i.e., sufficiently small $1 - W_2$) or when the direct positive effect on the beneficiary species 1 is large enough (sufficiently large $1 - W_1$). Moreover, a net negative effect is more likely when the benefactor species disproportionally increases (or does not decrease) the colonization rate of the beneficiary species 1 to make the trait-mediated indirect effect negative. Interestingly, even when benefactor's direct effect on the beneficiary species 2 is positive and that on the beneficiary species 1 is negative, a net negative effect can occur when the benefactor increases the competitive superiority of the beneficiary species 1 by changing the colonization rates of the two beneficiary species (Fig. 5.3c), underscoring the potential importance of trait-mediated indirect effects.

5.3.4 *Indirect Facilitation, Indirect Competition, and the Coexistence of Beneficiary Species*

Direct and indirect effects of benefactor species can affect the coexistence of beneficiary species. I examine the coexistence of two beneficiary species 1 and 2 in the presence of a benefactor species. I apply cases IA, IB, IIA, and IIB to compare different modes of competition-facilitation transition along the stress gradient. In case IA, the direct effect of benefactors changes from competition to facilitation with intensifying stress and transition occurs at the same stress levels for species 1 and 2 (Fig. 5.1a, b). At low stress levels (high λ and/or low σ), the direct competitive effect of the benefactor species suppresses the beneficiary species 1 (Fig. 5.4a, b). This invokes indirect facilitation of the beneficiary species 2, and permits its coexistence with the beneficiary species 1. At intermediate to high stress levels (intermediate λ and/or intermediate σ), facilitation by the benefactor species allows the beneficiary species 1 to persist, permitting its coexistence with the beneficiary species 2 (Fig. 5.4a, b). At even higher stress levels, the beneficiary species 1 does not persist but facilitation can support the persistence of the beneficiary species 2 up to the stress levels at which the benefactor species decreases and facilitation dwindles (Fig. 5.4a, b). Across the entire stress gradients, the net effect of the benefactor on the beneficiary species 2 changes from positive to negative at an intermediate stress level, and goes back to positive. The benefactor species tends to shift or extend the $\lambda - \sigma$ parameter region for coexistence toward higher environmental severity, and can substantially enlarge when indirect facilitation is sufficiently strong to promote coexistence in benign environments (Fig. 5.4c).

In case IB, the direct effect of the benefactor changes from negative to positive with increasing levels of stress as in case IA, but the stress-intolerant superior competitor (the beneficiary species 1) starts receiving the positive effect at less severe stress levels than the stress-tolerant inferior competitor (the beneficiary species 2) (Fig. 5.1c, d). This favors the superior competitor at intermediate stress levels and invokes indirect competition on the inferior competitor, making it difficult for them to coexist in benign environments (Fig. 5.4d, e). The benefactor species shift the $\lambda - \sigma$ parameter region for coexistence toward higher environmental severity (Fig. 5.5b). However,

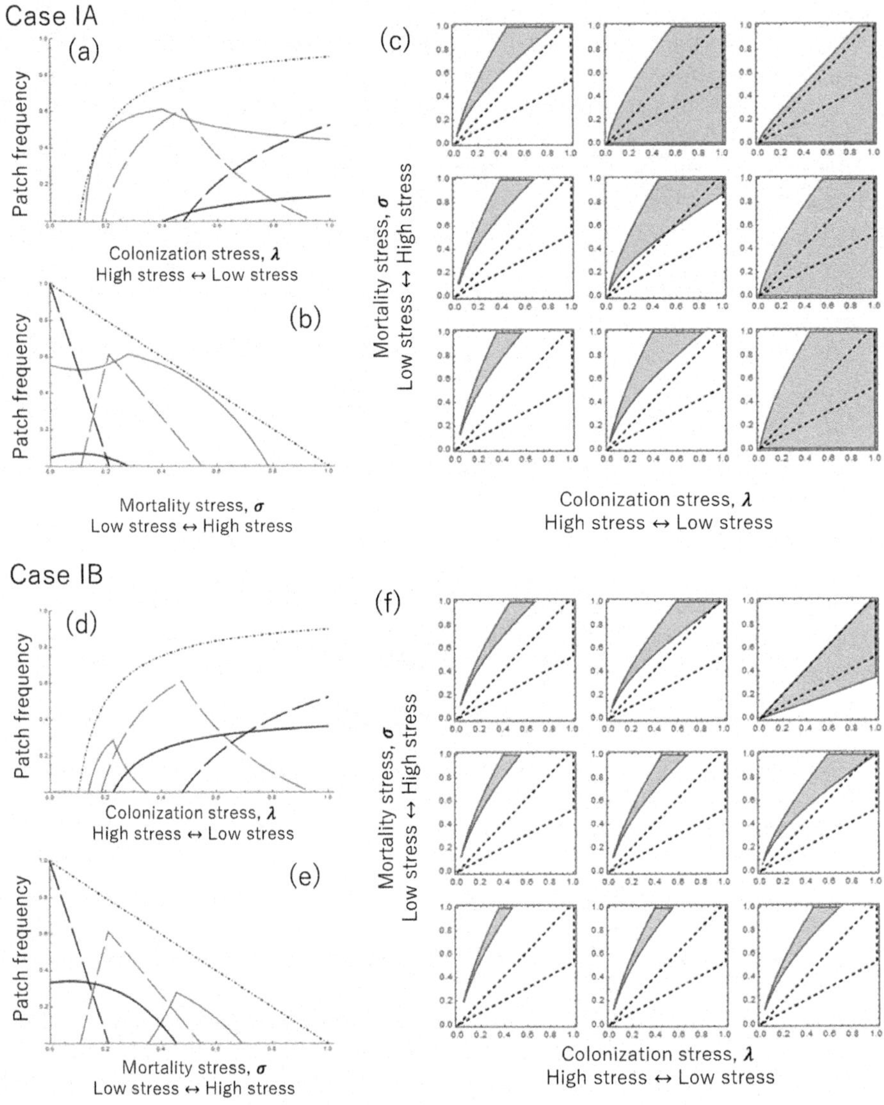

Fig. 5.4 Population abundances of a one-benefactor two-beneficiary system along the colonization (**a, d**) and mortality (**b, e**) stress gradients, and λ-σ parameter regions for the coexistence of the two beneficiary species (**c, f**). Case IA is shown in (**a–c**), and IB in (**d–f**). In (**a, b, d, e**), dot-dashed lines are benefactor patch frequencies. Black lines are the patch frequencies of the stress-intolerant superior competitor (the beneficiary species 1, $j = 1$), and gray lines are those of the stress-tolerant inferior competitor (the beneficiary species 2, $j = 20$). Dashed lines are beneficiary patch frequencies without the benefactor, and solid lines are the beneficiary patch frequencies with the benefactor. In (**c, f**), the regions surrounded by dashed lines is the coexistence parameter regions without the benefactor, and the gray regions are those with the benefactor. Parameter values are $k^u = 0.5$ and $k^v = 0.5$ in (**a, b**), $k^u = 0.6$ and $k^v = 0.4$ in (**d, e**); $k^u = 0.5$ (top-row panels), 0.7 (middle-row panels), 0.9 (bottom-row panels), and $k^v = 0.1$ (left-column panels), 0.3 (middle-

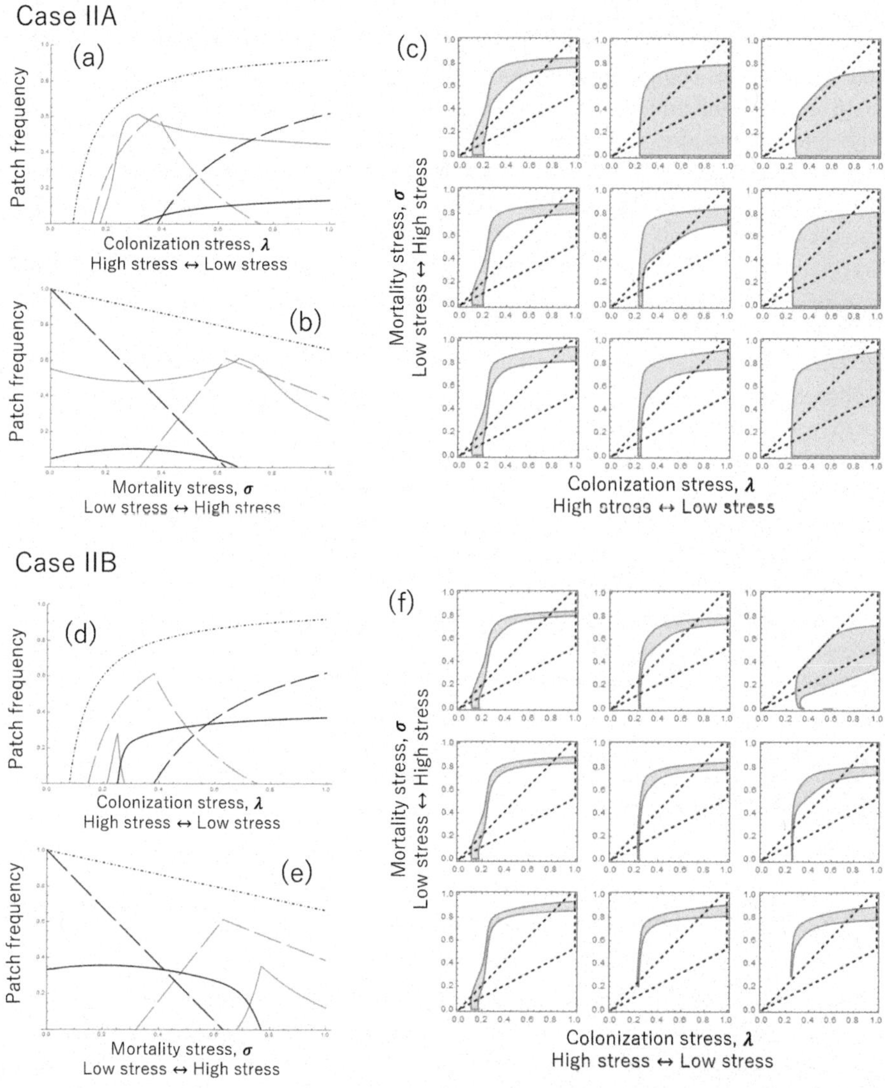

Fig. 5.5 Population abundances of a one-benefactor two-beneficiary system along the colonization (**a**, **d**) and mortality (**b**, **e**) stress gradients, and λ-σ parameter regions for the coexistence of the two beneficiary species (C, F). Case IIA is shown in (**a–c**), and IIB in (**d–f**). Parameter values are $\sigma = 0.4$ in (**a**, **d**), and $\lambda = 0.6$ in (**b**, **e**). Other parameter values and specifications are the same as in Fig. 5.4

Fig. 5.4 (continued) column panels), 0.5 (right-column panels) in (**c**, **f**) ; $\sigma = 0.5$ in (**a**, **d**), and $\lambda = 0.2$ in (**b**, **e**); $U = 1$ and $V = 0.2$ for all (**a–f**). Other parameter values are the same as in Fig. 5.1

unlike case IA, stronger direct facilitation on the superior competitor at intermediate stress levels causes indirect competition on the inferior competitor and precludes their coexistence from benign environments (Fig. 5.5b).

In case IIA, the direct effect of the benefactor transitions from negative to positive with intensifying stress, but shifts back to negative in extremely stressful environments (Fig. 5.1e, f). In Fig. 5.5a, b, the benefactor suppresses directly the beneficiary species 1 in benign environments, and promotes them within a narrow intermediate range of stress levels; the direct negative effect on the beneficiary species 1 in severest environments is not observed because it does not persist at such high levels of stress. The beneficiary species 2 receives indirect facilitation at low to intermediate levels of stress, and indirect competition when the benefactor species enables the beneficiary species 1 to persist at some intermediate levels of stress. At slightly higher stress levels, the beneficiary species 1 goes extinct, and direct facilitation increases the beneficiary species 2. At even higher stress, where the direct effect of the benefactor turns back to negative, the beneficiary species 2 is suppressed. The effect of the benefactor species on the λ-σ coexistence region is similar to case IA, favoring coexistence at higher environmental severity than without the benefactor (Fig. 5.5c). However, its direct negative effect eliminates the coexistence region from the most stressful environments (Fig. 5.5c).

In case IIB, transition between positive and negative direct effects of the benefactor species is the same as in case IIA, but the transition from negative to positive with increasing stress levels occurs at lower stress levels for the superior competitor (Fig. 5.1g, h). This promotes benefactor's indirect competition on the beneficiary species 2 mediated by the beneficiary species 1, and hampers their coexistence in benign environments (Fig. 5.5d, e, f). This parallels the difference observed between cases IA and IB.

5.3.5 *Species Diversity Along the Stress Gradients*

I then explore how facilitative and competitive effects of benefactor species along the stress gradients affect the species diversity of a multi-species community of beneficiary species. I assume a species pool of 20 beneficiary species. At a given level of stress, a subset of these species coexist at equilibrium, and I use the Shannon–Wiener index to evaluate species diversity. I examine cases IA, IB, IIA, and IIB with different competition-facilitation transition patterns along the stress gradient. Without the benefactor species, species diversity shows a unimodal pattern along the stress gradients with a peak at intermediate stress levels where stress-intolerant superior competitors and stress-tolerant inferior competitors coexist (Fig. 5.6a). In case IA, where increasing stress changes benefactor's effect from competitive to facilitative, the benefactor species generally increases species diversity across the entire stress gradients as compared to the case without the benefactor (Fig. 5.6b). This can be understood from the above result of two beneficiary species coexistence. In a many species community, indirect facilitation on stress-tolerant

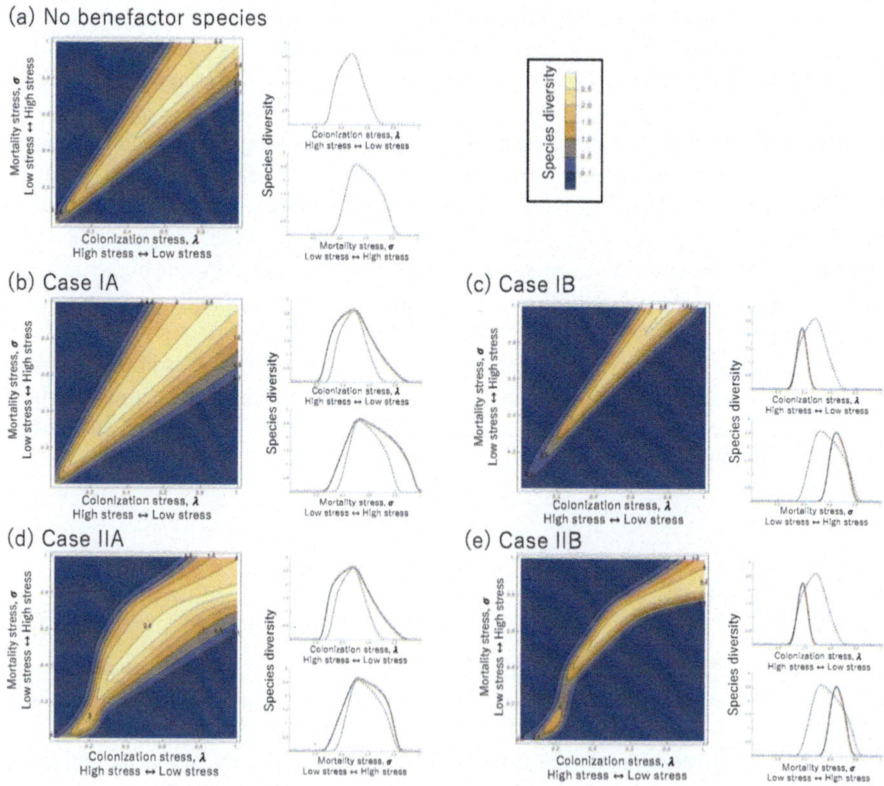

Fig. 5.6 Species diversity patterns across the colonization and mortality stress gradients. (**a**) With no benefactor species, and (**b–e**) with a single benefactor species. In each of (**a–e**), species diversity contours are plotted in panels to the left, and cross sections of species diversity responses along a single stress gradient at a given stress level of the other stress ($\sigma = 0.5$ or $\lambda = 0.5$) are shown to the right. In these panels, dashed lines show species diversity responses without a benefactor; solid lines show those with the benefactor. Parameter values are $k^{\mu} = 0.5$ and $k^{\nu} = 0.5$ in (**b, d**), $k^{\mu} = 0.6$ and $k^{\nu} = 0.4$ in (**c, e**). Other parameter values are the same as in Fig. 5.1. The diversity index is calculated from equilibrium abundances. The indices at similar stress levels show small variation due to species turnover. This variation is smoothed out with Gaussian filters when plotting the indices along the stress gradients

inferior competitors promotes species diversity at low stress levels, whereas direct facilitation on stress-intolerant superior competitors increases species diversity at high stress levels.

In case IB, in which benefactor's direct effect changes from negative to positive across the stress gradient but superior competitors start receiving facilitation at lower stress levels, the benefactor species shifts the peak of species diversity along the stress gradients toward severer stress levels (Fig. 5.6c). This shift occurs because facilitation preferring superior competitors at low stress levels strengthens indirect competition of the benefactor on inferior competitors of beneficiary species, hampering their coexistence.

In case IIA, the benefactor species has a negative competitive effect on beneficiary species in severest environments. This shifts the peak species diversity at severest colonization stress (low λ) and at severest mortality stress (high σ) toward less severe environments (Fig. 5.6d). This is probably because benefactor's direct negative effect on superior competitors at high colonization stress or high mortality stress causes indirect facilitation of inferior competitors, and promotes their coexistence.

In case IIB, facilitative effect of the benefactor differentially favors stress-intolerant superior competitors as in case IB. This strengthens indirect competition on stress-tolerant inferior competitors, lowering species diversity in low stress environments (Fig. 5.6e).

5.4 Discussion

I developed a framework of mathematical models to analyze the effects of competitive and facilitative effects of benefactor species on the population abundance and community properties of beneficiary species. This framework has four major strengths. (1) Empirical research suggests that there are diverse modes of transition of individual-level species interactions between competition and facilitation along the stress gradients. The current framework can implement such different transition modes into population-dynamic models of beneficiary species. (2) The framework considers the population dynamics of benefactor as well as beneficiary species. The prerequisite for benefactor species to affect beneficiary species is that their populations persist under given stress levels. Moreover, strength of their effects should depend on their population size. The current framework incorporates the mechanism through which stress levels determine the persistence and population size of benefactor species, control the strength of interactions between benefactor and beneficiary species, and thereby affect the population and community dynamics of beneficiary species. (3) The framework illustrates possible mechanisms of density- and trait-mediated indirect effects of benefactor on beneficiary species. (4) The framework produces general predictions about competitive and facilitative effects of benefactor species on coexistence and species diversity in communities of beneficiary species. I discuss these five points in more details with potential future extensions.

5.4.1 Competition-Facilitation Transition Along the Stress Gradient

The original SGH by Bertness and Callaway (1994) predicts that facilitation, as compared to competition, becomes more important in more stressful environments. Field tests of this prediction have produced mixed results (Callaway 2007). Summarizing empirical results, Michalet et al. (2006) pointed out that facilitation may rather diminish at extremely severe environments. They also emphasized that stress levels at which transition from competition to facilitation occurs along the stress gradient may depend on life history strategies of beneficiary species. Types of stress may be another important factor regulating the transition of interactions (Kawai and Tokeshi 2007; Saccone et al. 2009). Michalet (2007) pointed out that stress types interact with life history strategies of benefactor and beneficiary species to determine whether benefactors cause negative or positive effects on beneficiaries along the stress gradients. More specifically, facilitation may be strengthened on stress-intolerant beneficiaries as non-resource stress becomes severer, whereas competition may be more intensified on resource-demanding beneficiaries with severer resource stress. Following this suggestion, Maestre et al. (2009) summarized predictions that consider how the transition between negative and positive interactions depend on life history of benefactor and beneficiary species and whether stress factors are resource (such as water) or non-resource (such as temperature). In particular, they suggested that facilitation is most likely to be expected at intermediate environmental severity when stress factors are the resource type and benefactor and beneficiary species have similar life history strategies.

The current framework provides a mathematical formulation that can implement these diverse patterns of competition-facilitation transition along the stress gradients. A strength of the current formulation might be the ease of parameterization; results from empirical tests could be used to parameterize the model in a straightforward manner. The mechanistic formulation of the framework can explicitly consider whether a benefactor individual co-occurs with a beneficiary individual in the same locality and can incorporate separately measured relative strength of competitive and facilitative effects. With this framework, results of field tests that control the presence/absence of benefactor individuals and measure the net/separate effects of positive and negative interactions (e.g., Graff and Aguiar 2017) could be used to inform model parameterization.

The current framework considers two types of stress factors: one affecting colonization rates (λ) and the other mortality rates (σ). One could consider that resource availability, such as water deficiency, defines colonization stress because resource availability can affect individual growth (biomass production) and thus propagule production. Thus, the parameter λ could be interpreted as representing resource stress. On the other hand, severity of non-resource stress factors, such as temperature, could be considered as defining mortality stress because such environmental severity will increase mortality. Therefore, the parameter σ could be interpreted as a non-resource stress factor. With these interpretations, one could relate the difference between cases IA and IB, or between IIA and IIB, to different combinations of life history strategies between benefactor and beneficiary species to formulate each of the competition-facilitation transitions along the gradients of resource and non-resource stress suggested in Maestre et al. (2009). However, it is also likely that resource availability influences mortality and severity of the non-resource environments control individual production. Therefore, care would be necessary when relating colonization stress and mortality stress in this model to resource vs. non-resource stress factors.

One factor that can affect the competition-facilitation transition along the stress gradient is the spatial scale at which effects of these interactions are realized (Michalet et al. 2006). Although the analysis in this article assumed competition and facilitation occur on the same spatial scale (within the same patch), it would be straightforward to extend the framework to build a spatially explicit patch occupancy model. Such a spatial explicit framework could be useful to consider scale-dependent competitive and facilitative interactions between benefactor on beneficiary species.

5.4.2 Importance of Benefactor Population Dynamics on Effective Facilitation and Competition

The stress gradient hypothesis focuses primarily on the frequency of positive or negative species interactions at the individual level. However, for these interactions to take effect in regulating population dynamics of beneficiary species, population persistence and abundance of benefactor species should also be taken into account. Michalet et al. (2006) suggested that facilitative effects may become weak at extremely severe environments, making it difficult for beneficiary species to persist. Such decline of facilitative effects can occur when per capita strength of net positive interactions becomes weak, and/or when benefactor populations dwindle in severer environments. The present results demonstrate this. For example, even when benefactor species have facilitative effects on beneficiary species, their impacts should be smaller if their abundance is low (Fig. 5.2).

This point has a direct relevance when facilitation is applied to ecosystem restoration (Gómez-Aparicio 2009) and if benefactor populations are expected to be self-sustainable. To address this issue, the current framework formulates population-dynamic models of benefactor species in addition to those of beneficiary species. Although the analysis presented in this article assumes only a single benefactor species, future analysis should take into account responses of multi-species benefactor communities along the stress gradients to fully evaluate the strength of facilitation and competition from multiple benefactor species.

5.4.3 *Indirect Facilitation and Indirect Competition*

Indirect effects have been well studied in the field and theoretically, between species belonging to different trophic guilds (e.g., Wootton 1994; Abrams et al. 1996; Takimoto et al. 2009), as well as those within the same guilds, such as among plant species (e.g., Levine 1999; Brooker et al. 2008; Xiao and Michalet 2013). The current framework provides theoretical examples of such indirect interactions between benefactor and beneficiary species within the same trophic levels. Indirect facilitation and indirect competition can occur in a simplest three-species situation with a donor benefactor species, a transmitter beneficiary species, and a recipient beneficiary species. For example, consider that the transmitter is a stress-intolerant competitive species, which may be facilitated by a donor benefactor under non-resource type stress, and that the recipient is a stress-tolerant inferior competitor, which may receive less facilitative effect. Then if the donor facilitates the transmitter and elevates its abundance, the indirect effect on the recipient can be negative. On the other hand, if the donor competitively suppresses the transmitter, it will indirectly facilitate the recipient species. These indirect competition and facilitation by benefactor species on beneficiary species are important to determine the coexistence of beneficiary species (Fig. 5.3).

In particular, stress levels at which a benefactor facilitates or suppresses a transmitting superior competitor determine systematically the outcome of indirect effects on the coexistence of beneficiary species. If competitive suppression by the donor species acts on the transmitter species at low stress levels where the transmitter (superior competitor) would exclude the donor (inferior competitor) without the indirect effect, indirect facilitation can promote their coexistence at low to intermediate levels of stress. In contrast, if the donor facilitates the transmitter more strongly than the recipient at moderate to high stress levels, the transmitter would in turn suppress, and could competitively exclude, the recipient. These mechanisms seem to suggest a general prediction about the importance of indirect facilitation and indirect competition along the stress gradient; indirect facilitation may be most important at low-moderate stress levels, and indirect competition at intermediate-high stress levels.

Furthermore, the colonization performance of a transmitter beneficiary species, in addition to its abundance, can mediate the indirect effects of a donor benefactor species. This is because the donor benefactor species can change the colonization rate of the transmitter beneficiary species, which then alters its competitive superiority to a recipient beneficiary species. Since the donor species can also change the colonization rate of the recipient species, transmitter's competitive superiority to the recipient species depends on relative sign and magnitude of donor's direct effects on transmitter's and recipient's colonization rates. In trophic community ecology, it has been well conceptualized and recognized that the density and traits of transmitters mediate indirect effects (Wootton 1994). The current finding suggests that the concepts of density- and trait-mediated indirect effects can be similarly applicable to indirect facilitation and competition among interacting benefactor and beneficiary species within the same trophic levels.

5.4.4 Species Diversity Along the Stress Gradients

The traditional Grime model describes how competition and environmental severity drive patterns of species diversity in plant communities (Grime 1973). The model considered that the roles of competition in structuring plant communities become weakened as environmental stress intensifies. This leads to the prediction that diversity is maximized at intermediate stress levels where mild competition permits the coexistence of competitive stress-intolerant species and stress-tolerant inferior competitors, whereas diversity is decreased at lower stress levels where competitive exclusion suppresses stress-tolerant species, and at higher stress levels where stress-intolerant species are excluded. Hacker and Gaines (1997) incorporated positive interactions among plant species into the Grime model, and predicted that facilitation can enhance species diversity at intermediate to high levels of environmental severity. Michalet et al. (2006) pointed out that facilitation is weakened in the severest environments, which lowered diversity in these environments. Simulation models by Xiao et al. (2009) and Le Bagousse-Pinguet et al. (2014) considered facilitation and its waning at extreme environmental severity, and demonstrated that facilitation shifts the peak of species richness toward higher environmental severity and rather lowers species richness in mild environments.

Following up these models, the current framework moves a step forward by incorporating the recent refinements of the SGH prediction (Michalet et al. 2006; Maestre et al. 2009; Holmgren and Scheffer 2010) to generate predictions on species diversity response to environmental severity gradients. The results confirm some of the previous predictions that facilitation can increase the species diversity of a community at high environmental severity (Hacker and Gaines 1997; Xiao et al. 2009; Le Bagousse-Pinguet et al. 2014), and the breakdown of facilitation limits species diversity in extremely severe environments (Michalet et al. 2006; Xiao et al. 2009; Le Bagousse-Pinguet et al. 2014). Moreover, the current framework produced a number of predictions distinct from the previous models. First, adding a benefactor species increased species diversity at milder environments (Fig. 5.6b). Although this may seem to contradict the prediction of Xiao et al. (2009) and Le Bagousse-Pinguet et al. (2014), this is because the current framework considered competitive, as well as facilitative, effects of benefactors and indirect facilitation of the benefactors through superior competitors on inferior competitors promoted their coexistence. Second, inclusion of a benefactor species can reduce species diversity throughout the stress gradients with its peak shifting toward higher stress levels (Fig. 5.6c, e). This is because stronger facilitation and weaker competition from a benefactor on stress-intolerant superior competitors (Fig. 5.1c, d, g, h) favored them more than stress-tolerant inferior competitors. This prediction seems parallel to the pattern found by

Xiao et al. (2009) and Le Bagousse-Pinguet et al. (2014) that the species diversity peak shifts toward the higher side of environmental severity.

These findings underscore the importance of indirect facilitation and indirect competition on species diversity patterns. Indirect facilitation, invoked by benefactor's direct negative effect on competitively superior beneficiaries, can promote competitively inferior beneficiaries and increases species diversity at low to moderate stress levels. On the other hand, indirect competition, caused by benefactors facilitating competitively superior beneficiaries, can suppress competitively inferior beneficiaries and reduce species diversity at intermediate to high stress levels. This effect is more pronounced when stress-intolerant competitively superior beneficiaries receive facilitative effects from benefactors at lower stress levels than stress-tolerant competitively inferior beneficiaries. This demonstrates that beneficiary's life history strategies can influence not only the SGH prediction (Choler et al. 2001; Liancourt et al. 2005; Michalet et al. 2006; Michalet 2007; Maestre et al. 2009) but also community-level responses of species diversity along stress gradients. Testing these theoretical predictions in the field may be promising for our better understanding of how individual-level facilitation and competition can scale up to generate community-level patterns.

Acknowledgements I am thankful of advice from an anonymous reviewer who provided many useful references. This work is supported by JSPS KAKENHI 17H03736.

References

Abrams PA, Menge BA, Mittelbach GG, Spiller DA, Yodzis P (1996) The role of indirect effects in food webs. In: Food webs. Springer, Boston, MA, pp 371–395

Bertness MD, Callaway R (1994) Positive interactions in communities. Trends Ecol Evol 9:191–193

Brooker RW, Maestre FT, Callaway RM, Lortie CL, Cavieres LA, Kunstler G, Liancourt P, Tielbörger K, Travis JMJ, Anthelme F, Armas C, Coll L, Corcket E, Delzon S, Forey E, Kikvidze Z, Olofsson J, Pugnaire F, Quiroz CL, Saccone P, Schivvers K, Seifan M, Touzard B, Michalet R (2008) Facilitation in plant communities: the past, the present, and the future. J Ecol 96:18–34

Bruno JF, Stachowicz JJ, Bertness MD (2003) Inclusion of facilitation into ecological theory. Trends Ecol Evol 18:119–125

Callaway RM (2007) Positive interactions and interdependence in plant communities. Springer, Cham. ISBN 978-1-4020-6224-7

Choler P, Michalet R, Callaway RM (2001) Facilitation and competition on gradients in alpine plant communities. Ecology 82:3295–3308

Gómez-Aparicio L (2009) The role of plant interactions in the restoration of degraded ecosystems: a meta-analysis across life-forms and ecosystems. J Ecol 97:1202–1214

Graff P, Aguiar MR (2017) Do species' strategies and type of stress predict net positive effects in an arid ecosystem? Ecology 98:794–806

Grime JP (1973) Competitive exclusion in herbaceous vegetation. Nature 242:344–347

Hacker SD, Gaines SD (1997) Some implications of direct positive interactions for community species diversity. Ecology 78:1990–2003

Hastings A (1980) Disturbance, coexistence, history, and competition for space. Theor Popul Biol 18:363–373
Holmgren M, Scheffer M (2010) Strong facilitation in mild environments: the stress gradient hypothesis revisited. J Ecol 98:1269–1275
Kawai T, Tokeshi M (2007) Testing the facilitation-competition paradigm under the stress-gradient hypothesis: decoupling multiple stress factors. Proc Biol Sci 274:2503–2508
Le Bagousse-Pinguet Y, Xiao S, Brooker RW, Gross N, Liancourt P, Straile D, Michalet R (2014) Facilitation displaces hotspots of diversity and allows communities to persist in heavily stressed and disturbed environments. J Veg Sci 25:66–76
Levine JM (1999) Indirect facilitation: evidence and predictions from a riparian community. Ecology 80:1762–1769
Liancourt P, Callaway RM, Michalet R (2005) Stress tolerance and competitive-response ability determine the outcome of biotic interactions. Ecology 86:1611–1618
Maestre FT, Callaway RM, Valladares F, Lortie CJ (2009) Refining the stress-gradient hypothesis for competition and facilitation in plant communities. J Ecol 97:199–205
Maestre FT, Cortina J (2004) Do positive interactions increase with abiotic stress? A test from a semi-arid steppe. Proc Biol Sci 271(Suppl 5):S331–S333
Maestre FT, Cortina J, Bautista S (2004) Mechanisms underlying the interaction between *Pinus halepensis* and the native late-successional shrub *Pistacia lentiscus* in a semi-arid plantation. Ecography (Cop) 27:776–786
Michalet R (2007) Highlighting the multiple drivers of change in interactions along stress gradients. New Phytol 173:3–6
Michalet R, Brooker RW, Cavieres LA, Kikvidze Z, Lortie CJ, Pugnaire FI, Valiente-Banuet A, Callaway RM (2006) Do biotic interactions shape both sides of the humped-back model of species richness in plant communities? Ecol Lett 9:767–773
Michalet R, Brooker RW, Lortie CJ, Maalouf J-P, Pugnaire FI (2015) Disentangling direct and indirect effects of a legume shrub on its understorey community. Oikos 124:1251–1262
O'Brien MJ, Pugnaire FI, Armas C, Rodríguez-Echeverría S, Schöb C (2017) The shift from plant-plant facilitation to competition under severe water deficit is spatially explicit. Ecol Evol 7:2441–2448
Qi M, Sun T, Xue S, Yang W, Shao D, Martínez-López J (2018) Competitive ability, stress tolerance and plant interactions along stress gradients. Ecology 99:848–857
Saccone P, Delzon S, Pagès J-P, Brun J-J, Michalet R (2009) The role of biotic interactions in altering tree seedling responses to an extreme climatic event. J Veg Sci 20:403–414
Scrosati RA (2017) Community-level facilitation by macroalgal foundation species peaks at an intermediate level of environmental stress. Algae 32:41–46
Takimoto G, Iwata T, Murakami M (2009) Timescale hierarchy determines the indirect effects of fluctuating subsidy inputs on in situ resources. Am Nat 173:200–211
Tielbörger K, Kadmon R (2000) Temporal environmental variation tips the balance between facilitation and interference in desert plants. Ecology 81:1544–1553
Tilman D (1994) Competition and biodiversity in spatially structured habitats. Ecology 75:2–16
Wootton JT (1994) The nature and consequences of indirect effects in ecological communities. Annu Rev Ecol Syst 25:443–466
Xiao S, Michalet R (2013) Do indirect interactions always contribute to net indirect facilitation? Ecol Model 268:1–8
Xiao S, Michalet R, Wang G, Chen S-Y (2009) The interplay between species' positive and negative interactions shapes the community biomass-species richness relationship. Oikos 118:1343–1348
Zhang G, Zhao W, Zhou H, Yang Q, Wang X (2018) Extreme drought stress shifts net facilitation to neutral interactions between shrubs and sub-canopy plants in an arid desert. Oikos 127:381–391

Chapter 6
Mathematical Modeling on Microbes and Their Roles in Community and Ecosystem: How to Handle Microbial Diversity in Modeling?

Takeshi Miki

Abstract This chapter provides a synthesis of major existing models and theories of microbial communities, with a focus on their often overlooked roles in the maintenance of biodiversity and the regulation of ecosystem processes. This chapter covers the topics from both terrestrial and aquatic ecosystems where microbes are one of the major players of biogeochemical cycling and also mediated interactions between other organisms such as plants. In terrestrial ecosystems, the competitive relationship between plants is largely modified by soil microbes with diverse functional groups (e.g. microbial decomposer, pathogenic, or mutualistic groups) so that microbe-mediated plant–soil interactions have the potential to revise the classical theory of competitive coexistence. In aquatic ecosystems, a microbial food web including bacteria, fungi, protozoa, and phages structures the whole carbon and nutrient cycle. One challenging aspect of modeling microbial community is how to deal with taxonomic and functional diversity of microbes. Major modeling frameworks with different levels of complexity and the comparison of their performance will be discussed. Finally, a new modeling direction integrating mathematical models with bioinformatics will be also proposed.

Keywords Biodiversity · Ecosystem Functioning · Model complexity

T. Miki (✉)
Institute of Oceanography, National Taiwan University, Taipei City, Taiwan

Research Center for Environmental Changes, Academia Sinica, Taipei City, Taiwan

Faculty of Advanced Science and Technology, Ryukoku University, Otsu, Japan

A. Mougi (ed.), *Diversity of Functional Traits and Interactions*, Theoretical Biology,
https://doi.org/10.1007/978-981-15-7953-0_6

6.1 Microbes: Neglected Components in Ecosystems

Microorganisms have been used for one of the most common model systems in ecology from the early twentieth century. Not a few fundamental mathematical theories in ecology have been developed on them, e.g., the theories of competition and chemostat (Hsu et al. 1977; Tilman 1980; Hsu et al. 1981; Smith and Waltman 1995), and prey–predator interactions and their coevolution (Lenski and Levin 1985), which then have helped to understand the systems of plants and animals. However, mathematical theories focusing on the roles of microbes in nature remain underexplored despite highest abundance and second largest biomass (Bar-on et al. 2018) on Earth, which prevents the theoretical development of both general ecology and microbial ecology. In this sense, microorganisms had been under represented in ecology. However, from the end of the twentieth century to the early twenty-first century, the movement started to unifying microbial ecology and general ecology (Horner-Devine and Bohannan 2006). Along with this movement from empirical approach, there is substantial development of mathematical models and theories, which have contributed to the "unified ecology." Here, I can focus on a part of major topics of mathematical modeling of microbes at community and ecosystem scales only, especially focusing on bacteria. More extensive and systematic theoretical studies on viruses in community and ecosystems scales can be found elsewhere (Weitz 2016). Theories at the level of organismal and evolutionary scales and microbes as disease agents are out of focus here.

This chapter has three sections. The first section introduces core models to address how interactions of plants, soil, and microbes act as key drivers of community and ecosystem dynamics in terrestrial ecosystems. In addition, through using plant–soil–microbe interaction models as the example, two approaches to mathematically analyze relatively complex ecological models are argued: subsystems approach and dry experimental approach. The second section introduces two basic models to describe microbial interactions in lakes and oceans. One is the model for interactions between multiple species of heterotrophic bacteria and two different size classes of organic carbon (particulate organic carbon [POC] and dissolved organic carbon [DOC]). The other is the model for interactions between heterotrophic bacteria, its predator (heterotrophic nanoflagellate, HNF), and its parasite (bacteriophage). These models tell how microbial interactions affect nutrient and carbon cycling in aquatic ecosystems.

Accumulating evidence in the last two decades suggests the prevalence of the high taxonomic and functional diversities of microbes and the diversity and complexity of interactions between them. In the final section, I focus on the modeling methodology for handling effectively such diversities and complexities. For this purpose, I will extend the two models for aquatic ecosystems that are presented in the second section. As the first example, I compare the behavior of multiple theoretical dynamical models (TDMs) with different complexities, all of which are developed for investigating the non-equilibrium relationship between bacterial community and carbon export in the ocean. As the second example, I also compare the behavior of a

TDM with high complexity and that of a steady state model (SSM), both of which are intended to explain the impact of bacteriophage on bacterial carbon consumption and respiration. With these two examples, I would like to leave an open question on modeling approach for highly diverse and complex microbial systems, which would be also shared as a question of importance in modeling terrestrial ecosystems.

6.2 Terrestrial Ecosystems: Feedbacks Between Organisms and Environment

Basis of terrestrial ecosystems consists of plants and soil microbes, each of which primarily acts as producer and decomposer of organic materials, respectively, and thus are major drivers of material and energy cycling. Only upon the stability of functioning of the plant–soil microbe subsystems, other subsystems such as insects, birds, and mammals as well as humans can maintain their populations and their complex food web structures. Bidirectional interactions between plants and soil environments were first addressed by early twentieth century's plant ecologist Tansley, who defined the feedback system of organisms (e.g., plants) and environment (soil and climate) as *ecosystem* (Tansley 1935). Despite this idea of great importance, under the influence of resource competition theory (Hsu et al. 1977; Tilman 1980) and a kind of unidirectional, environmental determinism in competition and natural selection, community ecologists until 1980s had tried to explain competitive outcome of plants and species composition by environmental conditions such as light, water, and nutrient conditions only. At the same time, however, from the early 1970s to 1980s, global scale feedback concept, Gaia, was proposed and developed by Lovelock and their colleagues (Lovelock 1972; Lovelock and Margulis 1974; Watson and Lovelock 1983), but it had never been incorporated into community ecology until 1990s. Berendse (1994) successfully scaled down Gaia concept into local interactions between plant community and soil conditions and demonstrated that plant traits responsible for nutrient remineralization (but these are not directly related to exploitative competition for light and nutrients) are determinants of plant competition through plant–soil feedback (PSF). Plant individuals, as well as plant population and community, affect (in other words, "cultivate") soil abiotic (e.g., nutrient availability) and biotic (e.g., microbial pathogens) conditions that in turn alter plant growth, competition, and thus community composition, forming PSF (van der Putten et al. 1993; Berendse 1994). From a broader point of view, plants are regarded as ecosystem engineers (Jones et al. 1994) and PSF can be considered as a model system for niche construction and related concepts such as eco-evolutionary feedbacks (Matthews et al. 2014).

6.2.1 *Basic Mathematical Models for Plant-Centered View of PSF*

In the first generation of PSF studies, the plant species identity and species-specific traits are considered as the major drivers of PSF; in other words, the plant-centered view of PSF was first developed. Theoretical studies in this generation have focused on two major components that are controlled by plants: soil root-habiting microbes and soil nutrients, both of which in turn determine plant growth and fitness. Here, I summarize two canonical models of plant-centered view of PSF and mathematical analyses on them: (1) "microbe-mediated" PSF and (2) "litter-mediated" PSF (Ke and Miki 2015 and reference therein for more diverse aspects). Both of the models are a natural extension of Lotka–Volterra competition model for two plant species (*plant A* and *plant B*), given by

$$\frac{dP_A}{dt} = r_A P_A \left(1 - \frac{P_A + c_B P_B}{T_P}\right), \tag{6.1}$$

$$\frac{dP_B}{dt} = r_B P_B \left(1 - \frac{c_A P_A + P_B}{T_P}\right). \tag{6.2}$$

Here, r_i represent intrinsic population growth rate for plant species i ($i = A$, or B), c_A and c_B represent interspecific impact of species A on species B and that of species B on species A, respectively, and T_P represents the carrying capacity for the plant community, which is common for both species.

A simple invasibility analysis using rare-invader's fitness (Chesson 2000) can predict the competitive consequences of two plant species, which are classified into four distinct phases as the same as the original Lotka–Volterra competition model.

6.2.1.1 Microbe-Mediated PSF

Plant cultivates local microbial communities surrounding plant roots (both beneficial and detrimental groups, e.g., mycorrhiza and pathogenic fungi), which in turn affect plant performance and thus plant coexistence (Bever 1997; Bever 2003; also see Janzen 1970). Bever's seminal models assume that different plant species cultivate the soil microbial community differently and that changes in the microbial community have differential impacts on different plant species (Bever 1997; see Fig. 6.1a for model diagram adopted from Bever 2003 and Ke and Miki 2015). The microbial community density associated with *plant A and plant B* is presented by N_A and N_B, respectively, noting that each of them is not necessarily a single microbial species population but can be a microbial community. The model assumes that the relative abundance of different soil microbial communities, $S_A = N_A/(N_A + N_B)$ and $S_B = N_B/(N_A + N_B)$, linearly affects the per capita growth of plant population. Therefore, the Lotka–Volterra competition model (Eqs. 6.1 and 6.2) is modified as follows:

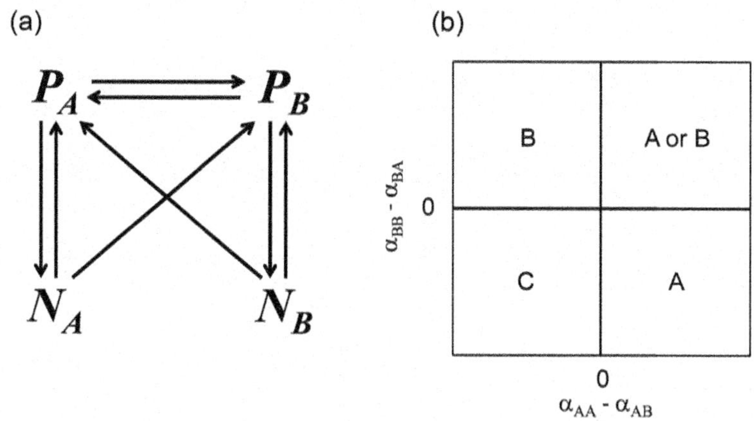

Fig. 6.1 Microbe-mediated PSF model. (**a**) Model diagram for microbe-mediated PSF model, derived from Bever (2003). (**b**) Consequences of microbe-mediated PSF on the equilibrium plant composition, characterized by PSF sign and strength $\phi_A = \alpha_{AA} - \alpha_{AB}$ and $\phi_B = \alpha_{BB} - \alpha_{BA}$. Plant sp. A or plant B monocultures are realized in region A or region B, respectively. Coexistence of two plant species is realized in region C, and alternative stable states of monocultures are realized in region A or B, depending on initial plant community composition. For the simplicity, c_A and c_B are set as 1.0 so that the boundaries for the distinct plant composition are exactly zero. (**c**) Consequences of litter-mediated PSF on the equilibrium plant composition, characterized by the litter decomposability of sp. A (e_A) and plant B (e_B). Plant sp.A or plant B monocultures are realized in region A or B, respectively. Coexistence of two plant species is realized in region C, and alternative stable states of monocultures are realized in region A or B, depending on initial plant community composition. The thresholds are $e_A^{**} = 0.156$ and $e_B^{**} = 0.120$, respectively, with the following parameter values: $T_P = 10.0$, $T_N = 20.0$, $r_A = 2.0$, $r_B = 1.0$, $K_A = 2.0$, $K_B = 0.3$, $m = 0.3$, $b = 1.0$

$$\frac{dP_A}{dt} = r_A P_A \left(1 + \alpha_{AA} S_A + \alpha_{BA} S_B - \frac{P_A + c_B P_B}{T_P}\right), \tag{6.3}$$

$$\frac{dP_B}{dt} = r_B P_B \left(1 + \alpha_{AB} S_A + \alpha_{BB} S_B - \frac{c_A P_A + P_B}{T_P}\right). \tag{6.4}$$

The parameters α_{ij} represent the sign and magnitude of PSF, i.e., how S_i influences P_j (i, j = A or B). Although this system is apparently driven by interactions between plants and microbes, species identity and species-specific traits are primary drivers of the system; each plant is assumed to cultivate species-specific microbial communities through species-specific traits.

Under the assumption of linear frequency (not density) dependency, the sizes of soil microbial communities increase linearly with the frequency of its host plant as

$$\frac{dN_A}{dt} = \frac{P_A}{P_A + P_B} N_A, \tag{6.5}$$

$$\frac{dN_B}{dt} = \nu \frac{P_B}{P_A + P_B} N_B, \tag{6.6}$$

where ν represents the relative influence of plant B on the microbial community compared to that of plant A. Since $S_A + S_B = 1$, the dynamics of frequency in different microbial communities can be derived by applying quotient rule of calculus to the time derivative of $S_A = N_A/(N_A + N_B)$ with Eqs. (6.5) and (6.6) as follows:

$$\frac{dS_A}{dt} = S_A(1 - S_A)\left(\frac{P_A}{P_A + P_B} - \nu \frac{P_B}{P_A + P_B}\right). \tag{6.7}$$

In order to classify the consequences of interspecific competition between two plants, mutual invasibility to monoculture equilibrium of either plant A or plant B (i.e., boundary equilibria of the system Eqs. 6.3, 6.4, and 6.7) can be checked. These equilibria are given as $\mathbf{E_A}\left(P_A^* = T_P(1 + \alpha_{AA}), P_B^* = 0, S_A^* = 1\right)$ and $\mathbf{E}_B\left(P_A^* = 0, P_B^* = T_P(1 + \alpha_{BB}), S_A^* = 0\right)$, respectively. The invasibility condition of plant A into $\mathbf{E_B}$ is evaluated by the positivity of the fitness of plant A as rare invader,

$$\lim_{P_A \to 0^+} \frac{1}{P_A} \frac{dP_A}{dt} \Big|_{\mathbf{E_B}} = r_A\left[1 + \alpha_{BA} - \frac{c_B T_P(1 + \alpha_{BB})}{T_P}\right] > 0. \tag{6.8}$$

In order to avoid mixed effects between PSF and direct competition between plants, we assume $c_A = c_B = 1$ (i.e., a neutral case such that two plant species are identical in terms of the spatial competition). Then, the invasibility criterion for plant A is written as $\alpha_{BB} - \alpha_{BA} < 0$.

This quantity, $\phi_B = \alpha_{BB} - \alpha_{BA}$, represents the sign and the strength of the impact of soil cultivation by species B on the invader fitness of species A. Such a community-level effect can be regarded as "community-level PSF," which is often different from the individual-level impact of soil cultivation by one species on the individual biomass growth of another species. Importantly, the sign of the community-level PSF is determined by the relative fitness of two species (which is equivalent to the sign of the invader fitness of rare invader since the fitness of the resident species at the demographic equilibrium is zero as the definition) and is not simply determined by the growth suppression or enhancement on each individual. It is worth noting that "negative" PSF is not necessarily caused by the negative impact of the microbial community cultivated with plant B on conspecific growth ($\alpha_{BB} < 0$) but is determined by the difference between the impact on conspecific (plant B) and heterospecific (plant A). For example, even with the positive impact on conspecific, e.g., with mutualistic relationship with mycorrhiza (i.e., $\alpha_{BB} > 0$), the community-level (net effect) PSF becomes negative and allows the invasion of heterospecific, if its positive impact is greater on heterospecific than conspecific ($\alpha_{BA} > \alpha_{BB} > 0$). Also, note that the effect of PSF and direct competition cannot be separable in an additive way if c_B is not equal to 1; the net strength and sign of the feedback is given by $\phi_B = c_B(1 + \alpha_{BB}) - \alpha_{BA} - 1$.

Similarly, the feedback strength and sign caused by plant A is obtained as $\phi_A = \alpha_{AA} - \alpha_{AB}$ from the fitness of plant B as a rare invader to $\mathbf{E_A}$ ($\lim_{P_B \to 0^+} ((1/P_B)dP_B/dt|_{\mathbf{E_A}}) > 0$). The parameter space of ϕ_A and ϕ_B can be partitioned into four regions based on these invasibility criteria (also confirmed by numerically solving ODEs, Fig. 6.1b). Coexistence (region C in Fig. 6.1b) is possible when both plant species cause negative community-level PSF (i.e., both ϕ_A and $\phi_B < 0$). This is comparable to the case when both plant species cause negative feedback in the pure Lotka–Volterra competition model ($c_A - 1$ and $c_B - 1 < 0$). Other combination of the sign of PSF results in either the dominance of single species (Region A or Region B) or bistability (Region A or B).

6.2.1.2 Litter-Mediated PSF

Plants have ability to alter nutrient cycling through species-specific plant litter traits (physico-chemical properties). Litter-mediated PSF model was proposed by Berendse (1994) and then extended by other studies (Miki and Kondoh 2002; Clark et al. 2005; Eppinga et al. 2011). Although microbes as decomposer are responsible for remineralization from plant litters, the first generation model of litter-mediated PSF keeps plant-centered views and assumes that rate of nutrient cycling is determined by plant traits (and environmental factors) only but not by microbial traits. Therefore, the model does not explicitly consider microbial population dynamics. Here, a simplified version of Berendse model (Ke and Miki 2015; Fig. 6.2a) is considered through modifying the intrinsic growth rate r_A and r_B in Lotka–Volterra model (Eqs. 6.1 and 6.2) as a function of soil nutrient N as,

$$\begin{aligned} dP_A/dt &= \text{(nutrient–dependent recruitement of new individuals)} \\ &\quad - \text{(loss by mortality)} \\ &= \frac{r_A N}{K_A + N}(1 - \frac{P_A + c_B P_B}{T_P})P_A - m_A P_A, \end{aligned} \tag{6.9}$$

$$\begin{aligned} dP_B/dt &= \text{(nutrient–dependent recruitement of new individuals)} \\ &\quad - \text{(loss by mortality)} \\ &= \frac{r_B N}{K_B + N}(1 - \frac{c_A P_A + P_B}{T_P})P_B - m_B P_B, \end{aligned} \tag{6.10}$$

where K_i (i = A or B) represents the half-saturation constant for plant species i (i = A or B) to uptake plant-available nutrient in the soil (N). The last term explicitly represents density-independent mortality with species-specific rate m_i. The resource dependent growth curves ($g_A(N) = r_A N/(K_A + N)$ and $g_A(N) = r_B N/(K_B + N)$) will cross at $\widehat{N} = (K_A r_B - K_B r_A)/(r_A - r_B) > 0$ when setting $r_A > r_B$ and $r_A/K_A < r_B/K_B$. This represents a trade-off between nutrient uptake strategies; the growth rate of plant A (or plant B) tends to be higher than the other species when $N > \widehat{N}$(or $N < \widehat{N}$) (Fig. 6.2b).

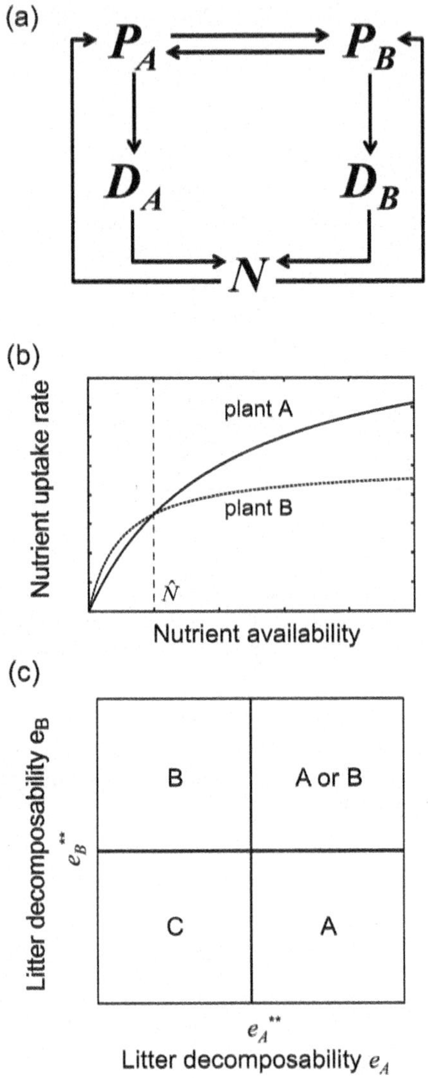

Fig. 6.2 Litter-mediated PSF model. (**a**) Model diagram for litter-mediated PSF model, derived from Berendse (1994) and Miki and Kondoh (2002). (**b**) The core trade-off relationship between plant A and plant B. The relative growth rate of plant A is higher than that of plant B when $N > \hat{N}$ while plant B has the growth advantage when $N < \hat{N}$

Dead plant materials (= plant litters) from plant A and plant B enter the plant-unavailable organic nutrient pool, with nutrient biomass denoted by D_A and D_B, respectively. With assuming a fixed amount of nutrient content per plant individual, b_i, the first term of Eqs. (6.11) and (6.12) represents the nutrient flux from plant populations to the organic nutrient pool,

$$\begin{aligned} dD_A/dt &= \text{(litter supply along plant mortality)} - \text{(loss by decomposition)} \\ &= b_A m_A P_A - e_A D_A, \end{aligned} \tag{6.11}$$

$$\begin{aligned} dD_B/dt &= \text{(litter supply along plant mortality)} - \text{(loss by decomposition)} \\ &= b_B m_B P_B - e_B D_B. \end{aligned} \tag{6.12}$$

The second term in Eqs. (6.11) and (6.12) represents the plant-centered view of litter-mediated PSF. Through neglecting the role of saprophytic microbes (microbial decomposers) and their taxonomic and functional diversities, specific decomposition rate, e_i, is temporally constant and plant-species specific. Remineralized nutrients are returned to the plant-available nutrient pool and taken up by plants for population growth. The dynamics of soil nutrient is given by,

$$\begin{aligned} dN/dt = &\text{(nutrient supply from remineralization of plant A litter)} \\ &+ \text{(nutrient supply from remineralization of plant B litter)} \\ &- \text{(loss by nutrient uptake by plant A)} \\ &- \text{(loss by nutrient uptake by plant B)} \\ = &e_A D_A + e_B D_B - \frac{b_A r_A N}{K_A + N}\left(1 - \frac{P_A + c_B P_B}{T_P}\right)P_A \\ &- \frac{b_B r_B N}{K_B + N}\left(1 - \frac{c_A P_A + P_B}{T_P}\right)P_B. \end{aligned} \tag{6.13}$$

Under this simple framework, the dynamical system was assumed to be closed in terms of nutrient mass balance, such that $b_A P_A(t) + b_A P_A(t) + D_A(t) + D_B(t) + N(t) = const. \equiv T_N$, where T_N represents the total nutrient content.

We set $c_A = c_B = 1$, $m_A = m_B = m$, and $b_A = b_B = b$ to examine the independent effects of litter decomposition rate, e_A and e_B as the determinant of PSF. The monoculture equilibrium of plant species i is given by,

$$\mathbf{E}_i\left(P_i^* = T_P\left(1 - m\frac{1}{r_i\left(N_{\mathrm{E}_i}^*\right)}\right), P_j^* = 0\right),$$

where $N_{\mathrm{E}_i}^*$ is the corresponding equilibrium for soil nutrient pool at the monoculture of plant species i and $r_i(N) = r_i N/(K_i + N)$. The invasibility of plant B into the monoculture of plant A is evaluated by the rare-invader fitness,

$$\lim_{P_B \to 0^+} \frac{1}{P_B}\frac{dP_B}{dt}\bigg|_{\mathrm{E_A}} = m\left(\frac{r_B\left(N_{\mathrm{E_A}}^*\right)}{r_A\left(N_{\mathrm{E_A}}^*\right)} - 1\right).$$

Then the invasibility criterion for plant B is written as $\phi_{\mathrm{A}} = 1 - \frac{r_B\left(N_{\mathrm{E_A}}^*\right)}{r_A\left(N_{\mathrm{E_A}}^*\right)} < 0$. This quantity ϕ_{A} can be regarded as the sign and strength of community-level PSF caused by plant A. The sign of this PSF is positive when $N_{\mathrm{E_A}}^* > \widehat{N}$. Similarly, the sign of strength of community-level PSF caused by plant B can be given as $\phi_{\mathrm{B}} = 1 - \frac{r_A\left(N_{\mathrm{E_B}}^*\right)}{r_B\left(N_{\mathrm{E_B}}^*\right)}$ and this is positive when $N_{\mathrm{E_B}}^* < \widehat{N}$.

The result from invasibility analysis can be further linked to the trade-offs of plant growth strategy and species-specific plant traits regarding nutrient recycling along the following steps. Note that $\widehat{N}$ is related to the trade-off of plant growth between two species. This is the critical level of nutrients above which (below which) plant A grows better than plant B (Fig. 6.2b). Then, when are these conditions realized? Intuitively, Plant A-cultivating soil nutrient level ($N^*_{\mathbf{E_A}}$) would be determined by the litter decomposability of plant A (e_A), but not by that of plant B (e_B). Besides, $N^*_{\mathbf{E_A}}$ should be an increasing function of litter decomposability e_A, and thus there should be a critical value of decomposability, $e_A{}^*$ that realizes $N^*_{\mathbf{E_A}} = \widehat{N}$, above which $N^*_{\mathbf{E_A}} > \widehat{N}$ is satisfied. Similarly, $N^*_{\mathbf{E_B}}$ is an increasing function of e_B, and thus there is a critical value of decomposability, $e_B{}^*$ that realizes $N^*_{\mathbf{E_B}} = \widehat{N}$, below which $N^*_{\mathbf{E_B}} < \widehat{N}$. This intuition is true, which can be demonstrated by mathematical analysis on the quadratic equation below (through setting r.h.s. of Eqs. 6.9–6.12 = 0 and substituting them into Eq. 6.13) for the equilibrium soil nutrient at the monoculture of plant species i,

$$f_i(N_i^*) \equiv (e_i r_i) N_i^{*2} + [(e_i + m)(r_i - m) b T_P - e_i r_i T_N] N_i^* - K_i b m T_P (e_i + m) = 0. \tag{6.14}$$

The above quadratic equation has unique positive root due to the positive quadratic coefficient with negative constant term. The PSF of plant A and that of plant B are positive when $f_A\left(\widehat{N}\right) < 0$ and $f_B\left(\widehat{N}\right) > 0$, respectively. These inequalities can be solved in terms of the litter decomposability as follows, respectively,

$$e_A > \frac{mbT_P r_A[(K_A r_B - K_B r_A) - m(K_A - K_B)]}{r_A(K_A r_B - K_B r_A)\left\{T_N - \left[(r_A - m) - \frac{(r_A - r_B)K_A m}{K_A r_B - K_B r_A}\right]\frac{bT_P}{r_A} - \frac{K_A r_B - K_B r_A}{r_A - r_B}\right\}} \equiv e_A^{**}, \tag{6.15}$$

$$e_B < \frac{mbT_P r_B[(K_A r_B - K_B r_A) - m(K_A - K_B)]}{r_B(K_A r_B - K_B r_A)\left\{T_N - \left[(r_B - m) - \frac{(r_A - r_B)K_A m}{K_A r_B - K_B r_A}\right]\frac{bT_P}{r_B} - \frac{K_A r_B - K_B r_A}{r_A - r_B}\right\}} \equiv e_B^{**}, \tag{6.16}$$

indicating that plant A can prevent the invasion of plant B when its litter decomposition rate is fast enough ($e_A > e_A{}^{**}$) while plant B can prevent the invasion of plant A when its litter decomposition rate is slow enough ($e_B < e_B{}^{**}$). When both conditions are satisfied, the system has alternative stable states of either $\mathbf{E_A}$ or $\mathbf{E_B}$ (Region A or B in Fig. 6.2c) while neither is satisfied the two plant species can mutually be invasible and can coexist (Region C in Fig. 6.2c).

These two lines of PSF models are the baseline for further development of PSF models. Here, the rare-invader fitness is proved to be a powerful mathematical tool to investigate the sign and magnitude of PSF and thus the consequences on plant coexistence. However, the analysis is limited to the equilibrium states of spatially homogeneous setting. Since PSF is in reality an important driver of temporal-spatial dynamics of plant communities, Chesson's modern competition theory framework, which is based on the rare-invader fitness but more focuses on temporal-spatial mechanisms of coexistence, can be applied to this system to better understand the roles of PSF in temporal (e.g., temporal storage) and spatial mechanisms of coexistence (Barabas et al. 2018).

6.2.2 *Advanced Mathematical Models for Plant–Microbe Interacting View of PSF*

In the second generation of PSF studies, especially in theoretical studies, soil microbes are not regarded as passive players that are conditioned by plants but as active players that modify the relationship between plant and soil. Figure 6.3 represents the conceptual framework of litter-and-microbe-mediated PSF. Here, two lines of mathematical models based on *plant-microbe-interacting view* of PSF are presented. Both of them combine microbe-mediated PSF and litter-mediated PSF but in different ways; one focuses on the interaction of microbial decomposer community (i.e., saprotrophic microbial community) and plant species-specific control on soil via litter decomposability while the other focuses on the interactions of rhizosphere microbial community (i.e., pathogens and mycorrhizal fungal

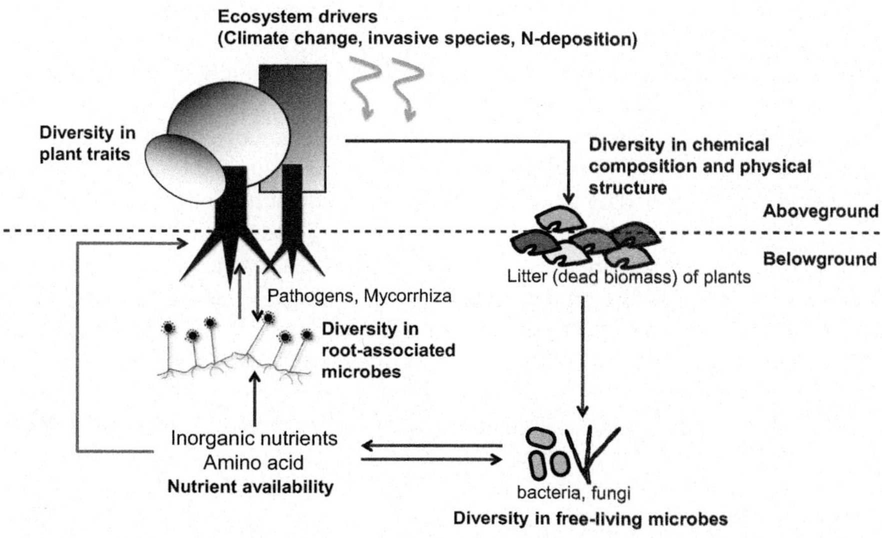

Fig. 6.3 Schematic diagram of litter-and-microbe-mediated PSF. Key components in the interaction between plants, microbes, and soil are shown

community) and plant species-specific control on soil via various traits including litter decomposability.

6.2.2.1 Litter-and-Saprotrophic-Microbe-Mediated PSF

The litter-mediated PSF model (Berendse 1994; Miki and Kondoh 2002) generates an inconvenient prediction, although these clearly demonstrate the importance of plant traits that are not directly related to exploitative resource competition but are related to nutrient recycling in plant competitive interactions. With the assumption of plant litter trait control on decomposition and nutrient remineralization processes, natural selection will favor the litter traits causing positive PSF and enhancing conspecific relative fitness, implying the prevalence of positive PSF within plant communities. The prevalence of positive PSF naturally predicts the difficulty in plant species coexistence in a local spatial scale, which is in good contrast to high local plant species diversity in natural environments. This inconsistency might imply the existence of certain neglected processes and players that should decouple or weaken plant litter traits and soil nutrient conditions. As such players, it is reasonable to focus on microbial decomposers such as bacteria and fungi and their diversity, since (1) recent culture-independent surveys reveal high taxonomic and functional diversity (Fierer et al. 2009; Strickland et al. 2009) and (2) a theory in general ecology on the relationship between biodiversity and ecosystem function tells that biodiversity has buffering effects against external forcing such as environmental fluctuations (Yachi and Loreau 1999).

In order to keep analytical tractability of a system, the model structure is as simple as possible (Fig. 6.4a). As the minimum levels of functional diversity of microbial decomposer, only two functional groups are considered: bacteria-type microbes (M_B) and fungi-type microbes (M_F). Similarly, the minimum level of chemical diversity of plant litter is considered: readily decomposable organic matter (D_R) and slowly decomposable organic matter (D_S). If plants produce D_R more than D_S as plant litter, it results in faster decomposition of plant litter into plant-available nutrient in the soil (N). The key idea of this model is that two microbial functional groups (M_B and M_F) compete for two resources (D_R and D_S) and thus microbial composition (i.e., the dominance of M_B, that of M_F or coexistence of them) and their relative abundance changes in the supply rate of D_R and D_S. The hypothesis here is that the flexibility of microbial composition and their relative abundance buffer the impact of changes in the litter chemistry (i.e., the fraction of D_R and D_S) on nutrient availability in the soil.

Here, in order to describe dependency of plant competition outcome on soil nutrient, a multispecies patch occupancy model (Tilman 1994) is used. The proportions of patches occupied by plant A (P_A) and plant B (P_B) are represented by Eqs. (6.17) and (6.18), as follows:

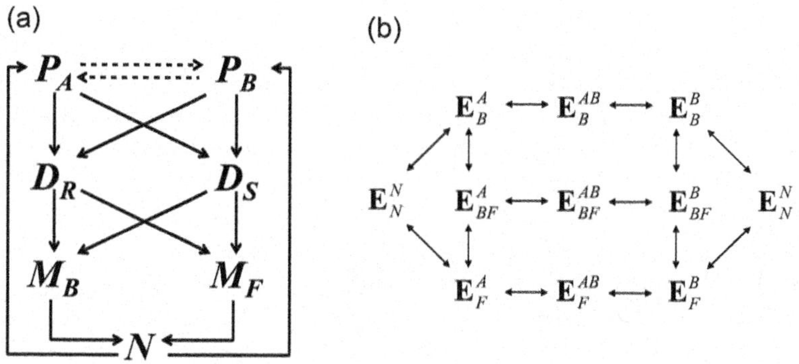

Fig. 6.4 Litter-and-saprotrophic-microbe-mediated PSF. (**a**) Flow diagram of the model. Nutrient flows are represented by solid arrows while competitive interactions for space between plant species are represented by dashed arrows. (**b**) The summary of compositional shifts of plants and microbes between different equilibria. The shifts represented by the arrows are predictable by mathematical analysis on invasibility and positivity of each equilibrium. Here, $\mathbf{E}_N^N$ represents the equilibrium where both plants and microbes cannot maintain their populations. Other equilibria are defined in the main text

$$\begin{aligned} dP_A/dt &= (\text{colonization to empty patches}) - (\text{loss by natural mortality}) \\ &\quad + (\text{colonization to the patches occupied by plant B}) \\ &= r_A N P_A (1 - P_A - P_B) - m_P P_A + r_A N P_A P_B, \end{aligned} \tag{6.17}$$

$$\begin{aligned} dP_B/dt &= (\text{colonization to empty patches}) - (\text{loss by natural mortality}) \\ &\quad - (\text{replacement by plant A}) \\ &= r_B N P_B (1 - P_A - P_B) - m_P P_B - r_A N P_A P_B. \end{aligned} \tag{6.18}$$

The first term of Eqs. (6.17) and (6.18) represents new recruitment in empty patches that is soil nutrient dependent, the second term represents loss through a density-independent morality rate, and the last term represents the recruitment or loss resulting from inter-individual competition of plant A and plant B. As the trade-off between two species, plant A has better competitive ability for occupied patches (plant A can replace plant B's occupying patches while plant B cannot replace plant A) while the recruitment ability of plant B to empty patches is better than plant A ($r_A < r_B$). Although the mechanism for the trade-off is different from those of Eqs. (6.9) and (6.10), the same type of nutrient level-dependent species composition is also realized in this model. With the positivity conditions and invasibility analysis, it is easily demonstrated that when $m_P/r_B < N^* < m_P/r_A$ ($\equiv N_B^{**}$), plant B dominates the plant community, when $N_B^{**} < N^* < (r_B/r_A)(m_P/r_A)$($\equiv N_A^{**}$), plant A and plant B coexist, and when $N_A^{**} < N^*$, plant A dominates the plant community.

In order to satisfy mass balance within the system, several assumptions regarding to nutrient cycling are necessary. Each plant individual is assumed to take up nutrient

b from the soil nutrient pool (N) at the moment of recruitment to establish biomass (corresponding to primary production) and release it into the litter pool at the moment of mortality (litter production, see Miki et al. 2010 for more realistic formulation). In addition, plant litter consists of two components (D_R and D_S) and the fraction of D_R of plant A and plant B is set as f_A and f_B; the higher fraction of D_R results in faster decomposition rate of plant species. Therefore, f_A and f_B represent species-specific litter quality in this model while e_A and e_B represent species-specific litter quality in the litter-mediated PSF model (Eqs. 6.9–6.13).

Decomposition rate of D_R and D_S is determined by their chemical quality and microbe-specific affinity that are parameterized by four values (k_{RB}, k_{SB}, k_{RF}, and k_{RF}), as well as the common decomposition coefficient determined e.g. by temperature (c_D). Although there are many possibilities for the combination of these four values, I focus on one possibility only with two realistic assumptions. First, independent of microbial functional groups, the decomposition rate of D_R is faster than D_S, i.e., $k_{RB} > k_{SB}$ and $k_{RF} > k_{SF}$ (litter chemistry or plant trait effect). Second, bacteria-like group (M_B) is better competitor for D_R while fungi-like group (M_F) is better competitor for D_S, i.e., $k_{RB} > k_{RF}$ and $k_{SB} < k_{SF}$ (microbial trait effect). With this setting (i.e., $k_{SB} < k_{SF} < k_{RF} < k_{RB}$), higher fraction of D_R in litter supply reasonably leads to higher relative abundance of bacteria-like group as the consequence of exploitative competition between two microbial groups (see below). The dynamics of D_R and D_S can be written as,

$$\begin{aligned} dD_R/dt = {} & (\text{production of litter from plant A}) + (\text{production litter from plant B}) \\ & - (\text{decomposition}) = f_A b m_P P_A + f_B b(m_P + r_A N P_A) P_B \\ & - c_D(k_{RB} M_B + k_{RF} M_F) D_R, \end{aligned} \tag{6.19}$$

$$\frac{dD_S}{dt} = (1 - f_A) b m_P P_A + (1 - f_B) b(m_P + r_A N P_A) P_B - c_D(k_{SB} M_B + k_{SF} M_F) D_S, \tag{6.20}$$

where the first and second terms of Eqs. (6.19) and (6.20) represent the litter production from P_A and P_B, respectively, and the third term is the loss through microbe-mediated decomposition.

There are two alternative fates for organic nutrient in plant litter that is decomposed by microbes; the fraction e_M is assimilated into microbial biomass (=microbial immobilization) while the other fraction $(1 - e_M)$ is released into soil nutrient pool (= net mineralization). By following mass balance, the dynamics of microbial biomass M_j and soil nutrient are given by,

$$\begin{aligned} dM_j/dt &= \text{(biomass production by decomposing organic matters)} \\ &\quad - \text{(loss by natural mortality)} \\ &= e_M(k_{Rj}D_R + k_{Sj}D_S)M_j - m_{M_j}M_j \quad j = M \text{ and } F, \end{aligned} \tag{6.21}$$

$$\begin{aligned} dN/dt &= \text{(loss by uptake of plant A)} - \text{(loss by uptake of plant B)} \\ &\quad + (\text{remineralization by } M_B) + (\text{supply through } M_B \text{ mortality}) \\ &\quad + (\text{remineralization by } M_F) + (\text{supply through } M_F \text{ mortality}) \\ &= -br_A N P_A(1 - P_A) - br_B N P_B(1 - P_A - P_B) \\ &\quad + (1 - e_M)c_D(k_{RB}D_R + k_{SB}D_S)M_B + m_{MB}M_B \\ &\quad + (1 - e_M)c_D(k_{RF}D_R + k_{SF}D_S)M_F + m_{MF}M_F, \end{aligned} \tag{6.22}$$

where the first and second terms of Eq. (6.21) represent microbial biomass growth and biomass loss due to mortality, respectively. The first through four terms of Eq. (6.22) are the loss through plant uptake, and other terms are supply from microbial decomposition (net mineralization) and from microbial biomass loss. Finally note that the ecosystem is closed, such that the total amount of nutrients (T_N) is conserved ($bP_A + bP_B + D_R + D_S + M_B + M_F = T_N$).

Ecologists may complain about the *simplicity* of the model since realistic levels of diversity of soil microbes and that of organic chemicals are not considered in this model. However, at the same time, mathematicians would complain about the *complexity* of the equations instead. Mathematical ecologists tend to develop a much simpler model (such as Eqs. 6.1–6.6) that is completely mathematically tractable or a much more realistic model of which complexity acts as a good excuse for avoiding analytical approach but relying on numerical solution of the differential equations only. Mathematical analysis on a model with intermediate complexity such as this model is not easier than very simple or realistic models. There are two directions of mathematical (analytical) and numerical analyses for such systems: *subsystems approach and dry experimental approach.* The first direction is to focus on subsystems with lower complexity, more exactly speaking, the boundary equilibria where some of components (some species or groups) are zero and check the positivity of these equilibria and invasibility conditions. The second direction is to mimic the design of the existing experiments that correspond to the model system in order to compare the existing empirical patterns with model behaviors, or the possible design of the experiments that are able to test if the mode behavior is valid.

In the subsystems approach for this model, the equilibria with single plant species i and single microbial group j ($\mathbf{E}^i_j$) and those with single plant species and two microbial groups ($\mathbf{E}^i_{BF}$) are considered. The equilibria with single microbial group j ($\mathbf{E}^i_j$) is given by,

$$0 = N^{*2} + \left[b - T_N + \frac{m'_{M_j}}{c_D}\left(\frac{f_i}{k_{Rj}} + \frac{1-f_i}{k_{Sj}}\right) + \frac{m_P}{m'_{M_j}} b \right] N^* - \frac{bm_P}{r_i}\left(1 + \frac{m_P}{m'_{M_j}}\right) \equiv F_{ij}(N^*), \tag{6.23}$$

$$P_i^* = 1 - \frac{m_P}{r_i N^*}, D_R^* = \frac{m'_{M_j} f_i}{c_D k_{Rj}}, D_S^* = \frac{m'_{M_j}(1-f_i)}{c_D k_{Sj}}, M_j^* = \frac{bm_P}{m'_{M_j}}\left(1 - \frac{m_P}{r_i N^*}\right), \tag{6.24}$$

noting that Eq. (6.23) has the unique positive root. In addition, this derivation tells that the ratio of microbial mortality (m_M) and assimilation efficiency (e_M) (i.e., $m'_{Mj} = m_{M_j}/e_M$) is an important determinant of soil nutrient level (N^*) and thus nutrient cycling.

Many of the positivity and invasibility conditions are obtained as algebraic inequalities by the following steps and these are found to be the conditions that species-specific litter quality (f_A and f_B) should satisfy, although the exact inequalities are not shown here (see Supporting Information of Miki et al. 2010). With these algebraic inequalities with numerical values of parameters, one can obtain the numerical values of the threshold litter quality with which the positivity or invasibility are switched. In this sense, the analysis is the combination of analytical (symbolic) and numerical methods but noting that numerical solution of the ODE systems (Eqs. 6.17–6.22) is not necessary at least for the boundary equilibria.

The positivity condition for plant biomass (i.e., $N^* > m_P/r_i$) is calculated as $F_{ij}(m_P/r_i) < 0$ for $\mathbf{E}^i_j$. This condition is equivalent to $f_i > f^E_{ij}$, implying that the system will collapse when plant litter quality is too low. This is the unique condition for the positivity of the equilibrium of single plant with single microbe.

The condition for each plant species (plant A or plant B) to generate a positive PSF (i.e., $N_A{}^{**} < N^*$ or $N^* < N_B{}^{**}$) is obtained from $F_{Aj}(N_A{}^{**}) < 0$ or $F_{Bj}(N_B{}^{**}) > 0$, respectively, for the single microbe system. When both plant species cause a negative PSF, these plants are likely to coexist (through mutual invasibility) for a system with one or two microbial groups. The invasibility analysis is also applied to judge whether $\mathbf{E}^i_j$ is stable in terms of microbial community composition. D_R-preferring bacteria-type microbes can prevent the invasion of D_S-preferring fungi-type microbes ($\frac{1}{M_F}\frac{dM_F}{dt}\big|_{\mathbf{E}^i_B} < 0$) when a fraction f_i (= litter quality) is sufficiently large ($f^{**}_{F\to iB} < f_i$), which can be analytically derived from the equilibrium. Similarly, fungi-type microbes can prevent the invasion of bacteria-type microbes when the litter quality is sufficiently low ($f_i < f^{**}_{B\to iF}$).

The equilibria with two microbial groups j ($\mathbf{E}^i_{BF}$) is given by,

$$
\begin{aligned}
0 = N^{*2} + [b - T_N + \frac{(k_{RB} - k_{SB})m'_{MF} - (k_{RF} - k_{SF})m'_{MB}}{c_D(k_{RB}k_{SF} - k_{RF}k_{SB})} \\
+ m_P b\{\frac{(k_{SF} - k_{SB}) f_i}{k_{SF}m'_{MB} - k_{SB}m'_{MF}} + \frac{(k_{RB} - k_{RF})(1 - f_i)}{-k_{RF}m'_{MB} + k_{RB}m'_{MF}}\}]N^* \\
- \frac{bm_P}{r_i}[1 + m_P\{\frac{(k_{SF} - k_{SB}) f_i}{k_{SF}m'_{MB} - k_{SB}m'_{MF}} + \frac{(k_{RB} - k_{RF})(1 - f_i)}{-k_{RF}m'_{MB} + k_{RB}m'_{MF}}\}] \\
\equiv F_{iBF}(N^*)
\end{aligned}
\tag{6.25}
$$

$$
\begin{aligned}
P_i^* = 1 - \frac{m_P}{r_i N^*}, D_R^* = \frac{k_{SF}m'_{MB} - k_{SB}m'_{MF}}{c_D(k_{RB}k_{SF} - k_{RF}k_{SB})}, D_S^* = \frac{-k_{RF}m'_{MB} + k_{RB}m'_{MF}}{c_D(k_{RB}k_{SF} - k_{RF}k_{SB})}, \\
M_B^* = bm_P P_i^*[\frac{k_{SF} f_i}{k_{SF}m'_{MB} - k_{SB}m'_{MF}} - \frac{k_{RF}(1 - f_i)}{-k_{RF}m'_{MB} + k_{RB}m'_{MF}}], \\
M_S^* = bm_P P_i^*[-\frac{k_{SB} f_i}{k_{SF}m'_{MB} - k_{SB}m'_{MF}} + \frac{k_{RB}(1 - f_i)}{-k_{RF}m'_{MB} + k_{RB}m'_{MF}}],
\end{aligned}
\tag{6.26}
$$

noting that Eq. (6.25) has the unique positive root, as the same as the equilibrium with a single microbial group.

The positivity condition for plant biomass (i.e., $N^* > m_P/r_i$) is calculated as $F_{iBF}(m_P/r_i) < 0$ for $\mathbf{E}^i_{BF}$. There are multiple conditions for the positive of the equilibrium with two microbes. The condition for positivity of the plant, $F_{iBF}(m_P/r_i) < 0$, is equivalent to $\frac{(k_{RB}-k_{SB})m'_{MF}-(k_{RF}-k_{SF})m'_{MB}}{c_D(k_{RB}k_{SF}-k_{RF}k_{SB})} < T_N - \frac{m_P}{r_i}$. The condition $k_{SF}m'_{MB} - k_{SB}m'_{MF} > 0$ and $-k_{RF}m'_{MB} + k_{RB}m'_{MF} > 0$ should be satisfied for non-negative nutrient biomass in plant litters (D_R^*, $D_S^* \geq 0$). This condition is true because the denominator is positive given the trade-off between two microbial groups. For the coexistence of two microbial groups (the positivity of M_B and M_F), the inequality $f^{**}_{R\rightarrow iF} < f_i < f^{**}_{S\rightarrow iB}$ should be also satisfied, noting that the threshold values in $f^{**}_{B\rightarrow iF} < f_i < f^{**}_{F\rightarrow iB}$ are related to the invasibility of the second microbe in the equilibrium $\mathbf{E}^i_j$.

The condition for each plant species (plant A or plant B) to generate a positive PSF (i.e., $N_A^{**} < N^*$ or $N^* < N_B^{**}$) is obtained from $F_{iBF}(N_A^{**}) < 0$ or $F_{iBF}(N_B^{**}) > 0$ (Eq. 6.25). When both plant species cause a negative PSF, these plants are likely to coexist (through mutual invasibility) for a system with one or two microbial groups. With the invasibility analysis applied to judge whether $\mathbf{E}^i_j$ is stable in terms of microbial community composition (see the section above for $\mathbf{E}^i_j$), and with a reasonable situation ($0 < f^{**}_{B\rightarrow iF} < f^{**}_{F\rightarrow iB} < 1$), the coexistence condition is simply $f^{**}_{B\rightarrow iF} < f_i < f^{**}_{F\rightarrow iB}$, noting that this condition is equivalent to the nonzero biomass condition of two microbes in the equilibrium $\mathbf{E}^i_{BF}$. This relatively simple condition for microbial composition is valid only under the trade-off relationships between bacteria and fungi for two resources (i.e., $k_{SB} < k_{SF} < k_{RF} < k_{RB}$). When this trade-off should be relaxed, for example, when finer levels of quality difference between two organic compounds or finer levels of microbial functional diversity (e.g., two different strategies of fungi) are considered, richer behaviors are expected to be observed even in the *subsystem* with single plant species, e.g., the bistability in terms of microbial community composition. In such a case, the full system with two plant species and two microbial groups would have much more locally asymptotically stable equilibria since both of the plant species and microbial groups can cause positive feedbacks, leading to multistability of the system. This kind of generalization of the model remains unexplored at this moment.

Although it is mathematically interesting to derive the invasion conditions for these competing microbial groups in the subsystem equilibrium with two plant species with single microbial group, it is difficult to analytically derive simple conditions for this scenario. It is also difficult to analytically derive the internal equilibrium with two plants and two microbial groups. Therefore, positivity and local stability of them is checked by numerically solving the ODE system (Eqs. 6.17–6.22). In conclusion, the possible compositional shifts, which can be predicted by analytical conditions, are summarized as the diagram (Fig. 6.4b).

In the dry experimental approach for this model, the subsystems with the single plant species with one or two microbial groups, but not the system with two plant species, is the main target. This is because the motivation of this modeling is to test the core hypothesis that the flexibility of microbial composition and their relative abundance buffer the impact of changes the litter chemistry (i.e., the fraction of D_R and D_S) on nutrient availability in the soil, following that plant diversity is not involved in this hypothesis itself. Of course, the final goal of this modeling is to demonstrate that microbial functional diversity and flexibility of microbial functional composition facilitate plant coexistence (see the original paper, Miki et al. 2010). However, it is difficult to test this model behavior (model prediction), i.e., whether plant coexistence is facilitated or not, by a simple experiment with a shorter time scale than plant generations. Therefore, the comparison of the system with or without microbial functional diversity (single microbial group or two groups) is a reasonable experimental setting for this model.

By numerically solving $F_{Aj}(N^*) = 0$ and $F_{ABF}(N^*) = 0$ with the numerical values of the equilibria values (Eqs. 6.24 and 6.26), the plot of the litter quality f_A vs relative abundance of bacteria-type microbes (Fig. 6.5a) and that of the litter quality vs the realized soil nutrient pool size (Fig. 6.5b) can be drawn. These results demonstrate that increasing litter quality does not affect the soil nutrient level when bacteria-type and fungi-type coexist ($f^{**}_{B \to iF} < f_i < f^{**}_{F \to iB}$), which is coupling with the increase in the relative abundance of bacteria-type microbes. On the contrary, in the subsystem with single microbial group (bacteria or fungi), there is always positive relationship between litter quality and soil nutrient level. This is the clear theoretical pattern that supports the hypothesis above. Following the comparison method of the system with or without microbial diversity in this dry experimental approach, Ushio et al. (2013) designed the (wet) experiment of soil microcosms and compared the systems with bacteria only, those with fungi only, and those with bacteria and fungi coexistence in terms of soil respiration and enzyme activity as the indices for decomposition

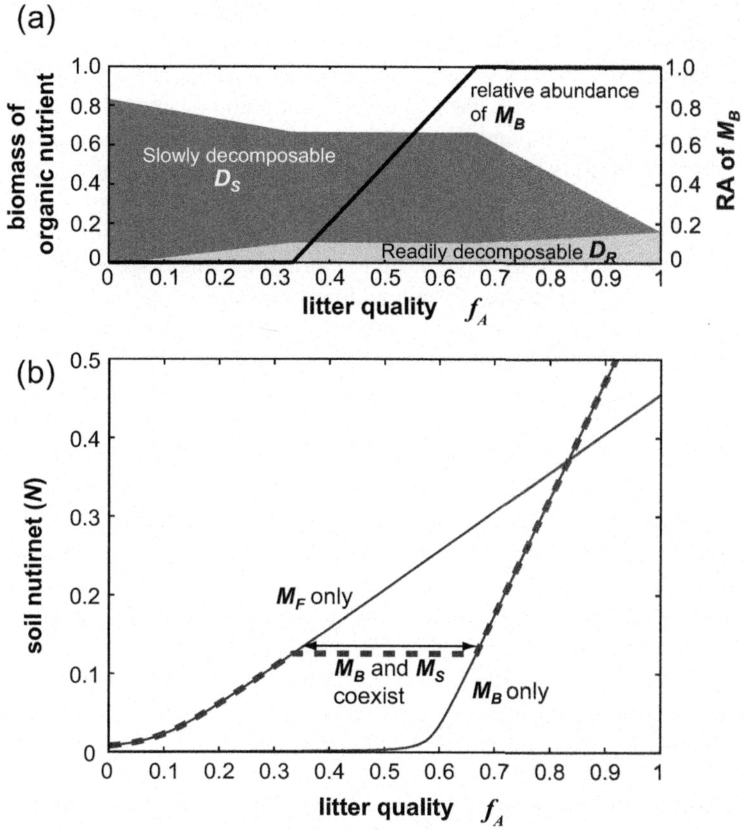

Fig. 6.5 Consequences of PSF on nutrient cycling. (**a**) Relationship between litter quality of sp. A (f_A), relative abundance (RA) of D_R-preferring bacteria-type microbes (M_B) at the equilibrium with plant sp. A only. (**b**) Relationship between litter quality and soil nutrient pool at equilibrium in systems with plant sp. A only. Solid lines represent the cases with the single microbial group (either M_B or M_F only). The dashed line represents the case where two microbial groups compete and the dominance of single group or their coexistence is realized depending on litter quality. Parameter values are $r_A = 5.0$, $r_N = 500.0$, $b = 0.2$, $m_P = 0.01$, $c_D = 1.0$, $k_{RB} = 2.0$, $k_{RF} = 0.5$, $k_{SB} = 0.1$, $k_{SF} = 0.2$, $e_M = 0.6$, $m_{MB} = 0.1$, $m_{MF} = 0.1$, and $T_N = 1.0$). This figure is modified from Miki et al. (2010)

activity, as well as microbial community composition. This microcosm experiments clearly demonstrated that (1) the microbial community composition changed most flexibly with litter quality and (2) the decomposition activity is least correlated with litter quality, in the bacterial–fungal coexisting systems.

6.2.2.2 Litter-and-Root-Associated-Microbes-Mediated PSF

There are two motivations to develop a model that considers the interactions between plants and root-associated microbes (pathogens and mycorrhizal fungi).

The primary motivation is to investigate the combined effects of litter-mediated PSF and microbe-mediated PSF and the secondary motivation is to challenge plant-centered trait-based approach in community and ecosystem ecology. In the Bever's microbe-mediated PSF model, the impact of plants on soil microbial composition and those of microbes on plant fitness are phenomenologically (or implicitly) formulated (Eqs. 6.3–6.6). Along with the first motivation, the mechanistic (or explicit) processes are also incorporated into the model. Pathogenic fungi and bacteria (hereafter, simply pathogens) are assumed to grow through exploiting and killing plants as resources but not directly related to nutrient cycling. Mycorrhizal fungi and plants are assumed to exchange carbon and nutrients based on the balance of their demand so that plant growth (maturation) and reproduction are modified not only by mycorrhizal fungi biomass but also soil nitrogen availability. These interactions between plants and pathogens, those between plants and mycorrhizal fungi are the major determinants of the relative abundance of these two functional groups of root-associated microbes. These interactions are formulated by the mechanistic processes (e.g., trophic interactions between plants and pathogens, and carbon and nutrient mass balance between plants and mycorrhizal fungi) in the model. In addition, the competitive interactions between pathogens and mycorrhizal fungi are also considered but phenomenologically formulated only by the Lotka–Volterra type competition. Along with the second motivation, both plants and microbial traits are considered such as their demographic parameters, plant litter quality, and carbon and nitrogen ratio of plants and microbes. In the trait-based ecology, especially in plant-centered trait-based approach, it tends to be assumed that there are key ecological traits as the determinant of population, community, or ecosystem properties. In the context of PSF, for example, litter traits have been considered as the determinant of the PSF strength. However, as is shown in the section above and in Miki et al. (2010), the impact of litter trait (quality) on soil nutrient, PSF sign, and thus plant competition depends on the presence/absence of functional diversity of microbial decomposers. In the same line, using this model, the relationship between plant traits and PSF strength is more extensively investigated. The core hypothesis is that the impact of plant traits on PSF strength highly depends on whether pathogens or mycorrhizal fungi are dominant in the root-associated microbial community.

In this section, the core assumptions of the model and the methods (subsystems approach and dry experimental approach) to investigate the model behavior are first presented. Then, a few model predictions and empirical tests are discussed while the mathematical formulations and mathematical analyses are not shown for page limitation (see Supporting Information of Ke et al. 2014).

The model consists of six state variables in an open ecosystem: seedlings (S) and adults (A) of a single plant species, pathogens (P), mycorrhizal fungi (M), organic plant litter (L), and soil inorganic nutrients (R) (Fig. 6.6a). Nutrients can be theoretically either nitrogen or phosphorus but nitrogen (N) is assumed for specific parameter value choice. S and A have the unit of density while other variables have the unit of nitrogen biomass. When considering nitrogen cycling, the densities of S and A are converted to biomass to satisfy mass balance. For both plants and microbes, carbon to nitrogen ratio is fixed. Two stages of plants (S and A) are considered because

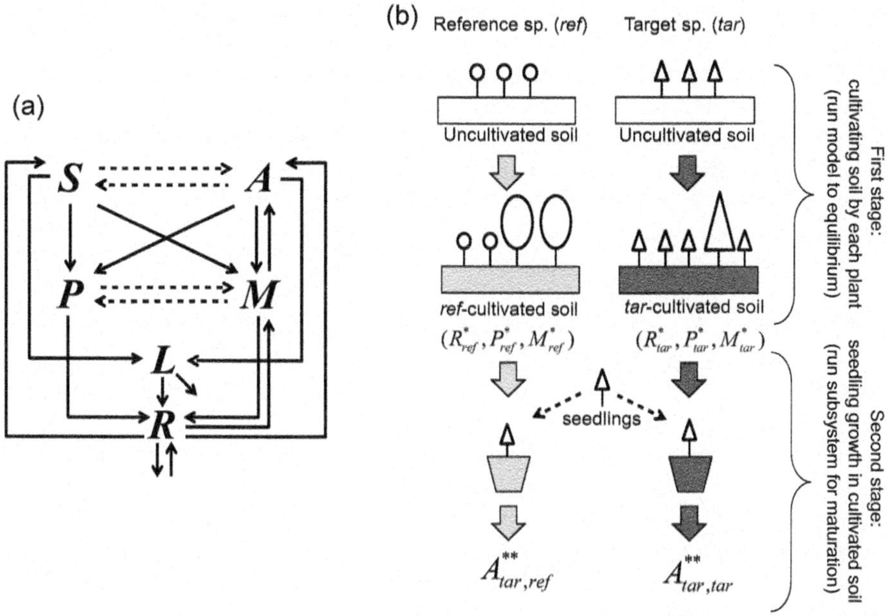

Fig. 6.6 Litter-and-root-associated-microbe-mediated PSF. (**a**) Model diagram representing interactions between six components. Solid arrows represent the fluxes of nutrient driven by trophic interactions, exchange of carbon and nitrogen, and nitrogen flux across the system boundary (deposition and leaching). Dashed arrows between *S* and *A* represent maturation and reproduction and nutrient fluxes along with these demographic processes, while those between *P* and *M* are direct competition between two microbial groups, resulting in their mortality and nutrient regeneration into *R*. (**b**) Schematic diagram of dry experimental approach mimicking PSF growth experiments. Note that the second stage uses the subsystems approach also. These diagrams are modified from Ke et al. (2014)

seedling and adults interact with *P* and *M* differently. Nitrogen flows between these six pools are presented by solid arrows in Fig. 6.6a.

The growth rate (maturation rate to adults) of seedlings increases with soil nitrogen (*R*) but decreases with adult density due to light competition. Nitrogen uptake by adult individuals is allocated to reproduction and tissue production. Due to the size difference between *S* and *A*, the contribution of *S* to the litter pool is via its mortality only while *A* contributes to *L* via continuous supply (coming from tissue production) and its mortality. The key assumption for plant growth and reproduction is that these are limited by nitrogen uptake and that the carbon assimilation is determined by nitrogen uptake. The rate of carbon assimilation should be explicitly considered because this is the determinant of the exchanges of nitrogen and carbon between plants and mycorrhizal fungi.

When plants interact with mycorrhizal fungi, plants increase carbon fixation and release the additional fixed carbon as root exudations for mycorrhizal fungi use while mycorrhizal fungi take up soil nitrogen and a part of it is transferred to plants in exchange for carbon products from plants. In order to avoid parasitic relationship between plants and mycorrhizal fungi and escalation of mutualistic exchange, it is assumed that carbon and nitrogen transfer between them is always positive and upper bounded. The realized fluxes between them are determined by the difference between maximum carbon supply from plants, and carbon demands of mycorrhizal fungi. Fungal carbon demands, to maintain the fixed C:N ratio in their biomass, are determined by soil nutrient availability and carbon supply from plants is upper bounded so that biomass growth of mycorrhizal fungi can be either C or N limit depending on the level of soil nutrient R.

The interactions between plants and pathogens are much simpler than those between plants and mycorrhizal fungi. Infection of pathogens results in plant mortality and a fraction of plant nitrogen content contributes to pathogen growth (assimilation) while unassimilated part of plant nitrogen is released into either organic litter pool (L) or inorganic nutrient pool (R).

There are several simple assumptions for nitrogen cycling. Litter decomposition is simply formulated by exponential decay in the same way as Eq. (6.11) without considering the dynamics of microbial decomposers. Nitrogen stored in pathogens and mycorrhizal fungi biomass is released directly into the soil nutrient pool. In this open ecosystem model, total nitrogen within the system is influenced by deposition (external input) and leaching (loss) of L and R.

As a dry experimental approach, the competition between multiple plant species is not explicitly considered in this model but a standard experimental setting is simulated, which intends to compare tree seedling growth response in conspecific- and heterospecific-cultured soils and to quantify PSF strength (Klironomos 2002; Brinkman et al. 2010; Mangan et al. 2010). This is because this simple experimental method is able to predict community properties. To mimic the experimental setting, two hypothetical plants are considered: a reference plant species (*ref*) and a target species (*tar*). The reference plant species have typical trait values calculated from empirical studies while one of the traits of the target species is different from the reference species.

The dry experimental approach consists of two stages (Fig. 6.6b). In the first stage, the different effects of *ref* and *tar* species on the soil environment are simulated by running the model ODEs numerically to equilibrium. The equilibrium values of soil N (R_k^*) and microbes ($P_k^*$and M_k^*) represent the soil properties cultivated by plant species k (k = *ref* or *tar*) and the difference of these values between *ref* and *tar* represents trait-specific effects on the soil environment. The second stage is designed to calculate the PSF strength for the target plant species and this stage uses a subsystems approach. In order to mimic short-term seedling growth evaluation in glasshouse conditions, the subsystem model considers seedling growth and mortality, which are influenced by mycorrhizal fungi and pathogens but do not consider adult reproduction and light competition between seedlings and adults. Adult mortality and litter dynamics are also neglected. The impact of soil conditions on plant seeding dynamics is temporally fixed with the equilibrium values (R_k^*, P_k^*, M_k^*). Under this constant soil environment that is determined by soil cultivation in the first stage, the subsystem model is numerically run until the density of seedlings approaches zero (through mortality or maturation to adults). At this time point, the abundance of adults of the target species ($A^{**}_{tar,k}$) represents the growth response of the target species under the soil cultivated by species k ($k = ref$ or tar). The PSF strength of target species is then determined by the log-ratio of growth response in two different soils as follows:

$$PSF_{tar} = \log\left(\frac{A^{**}_{tar,tar}}{A^{**}_{tar,ref}}\right). \tag{6.27}$$

Positive (or negative) PSF_{tar} indicates that the growth of the target plant species is promoted (or suppressed) in conspecific-cultivated soils due to deviation of the specific trait from the referent plant species. With the common degree of deviation from the reference values for each trait, the above analysis is repeated for all traits and the comparison of the magnitude of PSF_{tar} between different traits tells the relative trait importance. Since the relative trait importance as the determinant of PSF is expected to depend on microbial community composition, the same analysis is repeated under the following four scenarios considering: (1) litter-mediated PSF only without any root-associated microbes ($P = M = 0$), (2) litter-mediated PSF and pathogen-mediated PSF ($P > 0$, $M = 0$), (3) litter-mediated PSF and mycorrhiza-mediated PSF ($P = 0$, $M > 0$), and (4) litter-mediated PSF and microbial-mediated PSF driven by both pathogens and mycorrhizal fungi (P, $M > 0$). It is possible to investigate how root-associated microbes affect trait control on PSF strength by comparing the results across different scenarios.

With this modeling analysis, Ke et al. (2014) found that litter quality is not a major PSF determinant when pathogens are dominant in a root-associated microbial community. However, the impact of litter quality increases with the relative abundance of mycorrhizal fungi since nutrient supplied to plants via mycorrhizal fungi promotes plant growth and litter production, which in turn enhances the mycorrhiza growth and soil nutrient cycling. Another important prediction from this model is that plant C:N ratio is always a major determinant of PSF independently of microbial community composition since it affects both pathogen growth and the relationship between plants and mycorrhizal fungi.

Interestingly, the importance of litter quality in PSF is predicted again to depend on soil microbes (see the Sect. 6.2.2.1). However, since this prediction is relatively new and not popular in this field, many experimental studies did not prepare pathogen-only or mycorrhiza-only soil to quantify PSF strength. Only the results from Schnitzer et al. (2011) fit with the simulated settings to separate pathogens and mycorrhiza. Ke et al. (2014) reanalyzed the results of PSF strength with other publications for litter quality and demonstrated the positive relationship between litter quality and PSF strength in mycorrhiza-dominated soils but no statistically significant relationship in pathogen-dominated soils. This analysis clearly supports the model prediction.

6.2.3 Implication and Future Directions

Two distinct models above for plant–microbe interacting view of PSF clearly demonstrate the importance of microbial functional diversity and composition in predicting the impact of plant species-specific traits on PSF and thus community and ecosystem properties. This is a caution for the plant-centered view of trait-based ecology. Generally speaking, it is naïve to assume that traits of single species, single functional group, or single trophic level determine major ecological properties across scales. Therefore, it is not fruitful to switch from plant-centered view to microbe-centered view through addressing too much the importance of microbial diversity and composition (e.g., Toju et al. 2018) but is important to keep focusing on bidirectional relationships between plant and microbes (e.g., Mariotte et al. 2017) especially for understanding long-term dynamics of plant–soil systems in natural environments and for long-term sustainability in agroecosystems. In addition, even if the models above imply the importance of diversity and composition of microbial community in PSF, we should acknowledge that the drivers of these microbial community properties are unclear yet. If plant traits are major drivers of microbial diversity and composition, plant-centered view in PSF and plant trait-based ecology still survive and microbes can still be passive players in PSF (e.g., Eppinga et al. 2018). Only when microbial community properties are determined by the factors independently determined from plant traits, such as climate conditions, soil physical properties, and human disturbances, both of plants and microbes should be considered as active players in PSF. It is also possible that soil microbes govern everything and then the microbe-centered view will survive. It is too early to have consensus on this issue.

Plant–microbe interacting view of PSF is also applicable for understanding and forecasting plant community and ecosystem responses under climate change (Miki and Doi 2016) or other ecosystem drivers such as exotic species invasion (Miki 2012a). Incorporating this view into a more quantitative forest dynamic model will be one of the future directions (a prototype is presented by Miki and Doi 2016). From a more theoretical aspect, it is also important to extend the concept and methodology in PSF evaluation beyond pairwise competition perspective that basically relies on the rare-invader fitness as the index of PSF (Ke and Miki 2015). Some studies already focus on more complex plant competitive interactions (Lankau et al. 2011; Gallien et al. 2017; Levine et al. 2017; Saavedra et al. 2017) and this direction will be further explored.

6.3 Aquatic Ecosystems: Microbial Oceans

When were microbes, especially bacteria, first recognized as an important player in aquatic ecosystems? From environmental microbiology or microbial ecology point of view, Pomeroy is one of the first scientists to address the role of microbes (bacteria and protozoa) in energy pathway in marine food chains (Pomeroy 1974) and then Azam et al. (1983) proposed the concept of microbial loop as a core energy loop starting from dead organic carbon. However, in a broader sense, Lindeman (1942), who proposed the theory of trophic dynamics and the concept of trophic levels, already pointed out the importance of bacteria in "food-cycle" relationships in aquatic ecosystems, by placing the bacteria in the center of his diagram of food-cycle relationship (Fig. 1 in Lindeman 1942). Key concepts in material cycling in aquatic ecosystems are overviewed in Fig. 6.7. Although there are some differences between freshwater and marine water ecosystems in food web members and pathways of material cycling, the diagram in Fig. 6.7 consists of the members and pathways from lakes and oceans together. In pelagic regions of the lakes and oceans, phytoplankton is the key primary producer. The fixed carbon dioxide (CO_2) by phytoplankton is partly consumed by zooplankton. However, the remaining part of particles (particulate organic carbon, POC) is potentially sinking into the deeper layers of the systems, especially in deep oceans, these fractions of carbon are sequestrated up to 1000 years (Sarmiento and Gruber 2006). The vertical transport of carbon from the surface to deep layers is called biological pump, and the export efficiency of the biological pump (=proportion of export to total production)

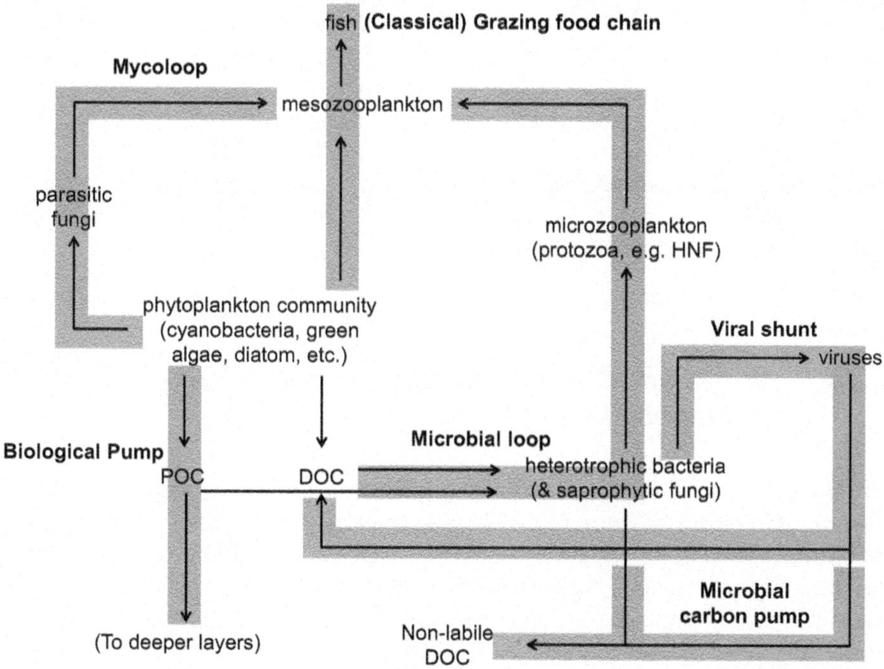

Fig. 6.7 Food web and carbon cycling in surface water of aquatic ecosystems. Schematic diagram represents carbon fluxes along trophic interactions as well as physical transport. Shaded pathways represent biological pump, classical grazing food chain, microbial loop, mycoloop, viral shunt, and microbial carbon pump

changes temporally and spatially (Ducklow et al. 2001). The export efficiency is influenced by phytoplankton community composition, zooplankton grazing, and activity of heterotrophic bacteria. At the same time, the other component, dissolved organic carbon (DOC) is too small for zooplankton and then potentially accumulated in the surface waters. Heterotrophic bacteria can convert both of DOC and POC into bacterial biomass and then transferred again into higher trophic levels by microbial food chain; this food-cycle is called microbial loop (Azam et al. 1983; Pomeroy et al. 2007). This recycling process of unused organic carbon is disturbed by bacteriophage (virus), which kills the host bacterial cells via lytic lifecycle, and less amount of organic carbon can be transferred to grazers of bacteria (protozoa); this is called viral shunt or viral loop (Fuhrman 1999; Suttle 2007). Both of the microbial loop and viral shunt are responsible for the fate of organic carbon produced by phytoplankton and determine the balance between bacterial carbon biomass production (carbon recycling) and bacterial respiration, which affects the efficiency of biological pump (Miki et al. 2008) and global carbon cycling (Kwon et al. 2009). Recently another process is also addressed as the key determinant of carbon sequestration in the oceans, which is called microbial carbon pump (Jiao and Zheng 2011). Bacteria, viruses, and other microbes in the microbial food web are responsible to chemically convert readily decomposable organic carbon into less degradable carbon. This microbial carbon pump makes the residence time of carbon in the ocean longer and thus contributes to the carbon sequestration as well as biological pump. In addition to viruses, aquatic fungi are also recognized as key parasites in aquatic food webs. In particular, parasitic fungi on large phytoplankton such as diatoms that are inedible for zooplankton due to their size can indirectly transfer materials in inedible large phytoplankton to zooplankton via trophic interactions between small and edible fungal zoospores and zooplankton. This is called mycoloop in freshwater ecosystems (Kagami et al. 2004, 2014). In summary, various microbes have different but important roles in aquatic food webs and ecosystem dynamics.

6.3.1 Basic Mathematical Models for Microbial Loop and Microbial Food Webs

In this section, two basic models are introduced for microbial loop and microbial food webs: a model for interactions between heterotrophic bacterial community and two distinct fractions of organic carbon (POC and DOC) and a model for interactions between organic carbon, bacteria, virus, and protozoa. The importance of mass balance is discussed as the key constraint of modeling microbial food webs. More systematic development of mathematical models for interactions between bacterial community and viral community and those for the functioning of viral loop can be learned in Thingstad (2000), Miki and Yamamura (2005a), and Weitz (2016). The modeling approach in mycoloop can be found in Miki et al. (2011) and Kagami et al. (2014). As far as I know, the modeling approach for microbial carbon pump remains underexplored.

6.3.1.1 Mathematical Model for Interactions between Bacterial Community, POC, and DOC

Phytoplankton community produces operationally distinct types of organic carbon: POC (particles) and DOC. Both are important resources for heterotrophic bacteria: free-living cells of bacteria can consume DOC while particle-attached (or particle-associated) cells can consume POC by hydrolyzing them into smaller and simpler molecules (i.e., DOC). Recent microbial community studies suggest the existence of three distinct types of bacterial strategies (Riemann and Winding 2001): (1) free-living specialists that consume DOC only, (2) particle specialists that consume POC only, and (3) generalists that consume both POC and DOC via switching between particle-attached and free-living states. During phytoplankton bloom, the community composition of phytoplankton shifts along time from small-size groups that preferentially produce DOC to large ones that preferentially produce POC (Dunne et al. 2005). Such compositional shifts of phytoplankton and changes in DOC:POC supply ratio to surrounding water are expected to be followed by the shift of bacterial community composition in terms of the three strategies mentioned above.

In order to reproduce such dynamical succession of bacterial community composition observed in natural environments (Riemann and Winding 2001), a simple theoretical dynamical model (TDM) is developed. The model consists of three bacterial strategies and DOC and POC but at the same time, free-living cells and particle-associated cells should be distinguished for POC specialists and generalists (Fig. 6.8a). Two fractions of carbon resources: POC and DOC are assumed, with carbon concentration C_P and C_D (mgC L^{-1}), respectively. DOC specialists are always at free-living state with the density F_D while POC specialists and generalists are at either free-living or particle-associated state, with the density F_P and A_P, and F_G and A_G, respectively. These bacterial-related variables have the unit of densities of cells or aggregates (L^{-1}). Between these seven components, three main processes are considered: (1) colonization of free-living POC specialist and generalist cells to particles, forming aggregates, (2) hydrolysis of POC and subsequent processes, and

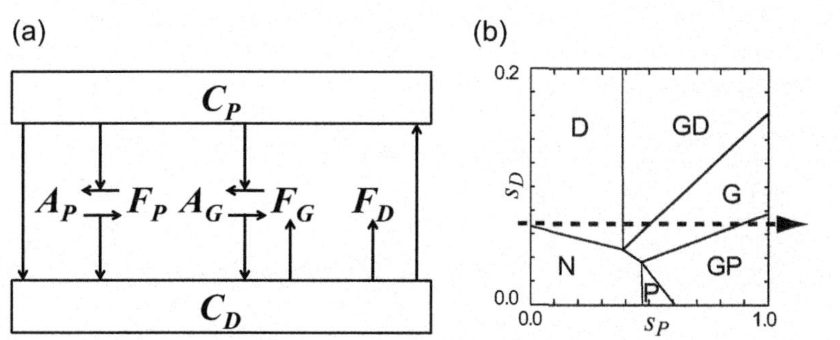

Fig. 6.8 POC-DOC-bacteria interaction model. (**a**) Model flow diagram. Arrows represent major fluxes of carbon in the model. Symbols are defined in the main text. (**b**) Relationship between carbon supply (s_P and s_D) and realized equilibrium bacterial composition. Parameter values are from Miki and Yamamura (2005b). These diagrams are modified from Miki and Yamamura (2005b)

(3) consumption of DOC by free-living generalists and DOC specialists. The material fluxes between these components are carbon-based and mass balance is carefully considered.

Bacterial Colonization to Particles To formulate colonization of free-living bacteria to particles (POC), the density of particles can be calculated using the carbon content per particle α, as C_P/α. Then the colonization rate of free-living cells of POC specialists and generalists can be formulated using clearance rate a_P and a_G, as $a_P(C_P/\alpha)$ and $a_G(C_P/\alpha)$, respectively. It is also simply assumed that each particle is colonized by n bacterial cells and then become aggregates with neglecting heterogeneity in the number of colonization between particles. It follows that the rate of aggregate formation is $a_P(C_P/\alpha)F_P/n$ and $a_G(C_P/\alpha)F_G/n$ for POC specialists and generalists, respectively. The carbon content of bacteria per cell is set as β; therefore, the carbon content per aggregate is calculated as $\alpha + n\beta$.

Hydrolysis of POC and Subsequent Processes For simplicity, the population dynamics of bacteria on each aggregate is not considered. It is assumed that bacterial cells on each particle consume it by hydrolysis (i.e., the conversion from POC to DOC) at rate h (i.e., the loss rate of aggregate [d^{-1}]). There are three alternative fates of hydrolyzed carbon; the fraction f_F is consumed by particle-associated bacteria for their production and respiration, the fraction f_D is released as DOC to the surrounding water, and the remaining fraction $(1 - f_F - f_D)$ is converted to non-labile forms so that this fraction is not considered for further processes. Since each aggregate disappears by hydrolysis, both of original n bacterial cells and newly produced cells $e_P\alpha f_F/\beta$ with the growth efficiency e_P become free-living cells. The unassimilated fraction of organic carbon $(1 - e_P)\alpha f_F$ is respired as CO_2.

Consumption of DOC It is assumed that the specific growth rates of free-living generalist and DOC specialist cells are $u_G C_D$ and $u_D C_D$ per cell (d^{-1}), respectively. Therefore, the loss rate of DOC can be calculated as $e_D^{-1} u_G C_D \beta F_G$ and $e_D^{-1} u_D C_D \beta F_D$, respectively, using the growth efficiency e_D. Unassimilated carbon is respired as CO_2.

With these major assumptions and some additional minor assumptions, the differential equations for this system can be obtained. The dynamics of aggregates with POC specialists and those with generalists are given by,

$$\begin{aligned} dA_i/dt &= \text{(formation of aggregates by bacterial colonization)} \\ &\quad - \text{(loss by hydrolysis)} \\ &\quad - \text{(loss due to bacterial mortality and vertical transport)} \\ &= a_i n^{-1} C_P \alpha^{-1} F_i - hA_i - (m_A + m_V)A_i,\ i = P \text{ and } G \end{aligned} \tag{6.28}$$

where m_A and m_V are loss rate of aggregates due to bacterial mortality (and returning to fresh POC pool) and loss due to vertical transport to the deeper layers (i.e., the complete loss from the system), respectively.

The dynamics of free-living cells for each of three strategies are given by,

$$
\begin{aligned}
dF_P/dt &= \text{(release from aggregates during hydrolysis)} \\
&\quad - \text{(loss by colonization to particles)} \\
&\quad - \text{(loss due to natural mortality)} \\
&= (n + e_P\alpha\beta^{-1} f_F)hA_P - a_P(C_P\alpha^{-1})F_P - m_F F_P,
\end{aligned} \tag{6.29}
$$

$$
\begin{aligned}
dF_G/dt &= \text{(release from aggregates during hydrolysis)} \\
&\quad - \text{(loss by colonization to particles)} \\
&\quad + \text{(replication by DOC uptake)} \\
&\quad - \text{(loss due to natural mortality)} \\
&= (n + e_P\alpha\beta^{-1} f_F)hA_G - a_G(C_P\alpha^{-1})F_G + u_G C_D F_G - m_F F_G,
\end{aligned} \tag{6.30}
$$

$$
\begin{aligned}
dF_D/dt &= \text{(replication by DOC uptake)} - \text{(loss due to natural mortality)} \\
&= u_D C_D F_D - m_F F_D,
\end{aligned} \tag{6.31}
$$

where m_F is the constant mortality per cell.

The dynamics of POC and DOC is given by,

$$
\begin{aligned}
dC_P/dt &= \text{(supply from primary production)} \\
&\quad - \text{(loss by colonization of bacteria, forming aggregates)} \\
&\quad - \text{(loss due to unspecified processes and vertical mixing)} \\
&\quad + \text{(conversion from DOC)} \\
&\quad + \text{(recovery of particles due to mortality of attached bacteria)} \\
&\quad + \text{(regeneration during bacterial mortality)} \\
&= s_P - n^{-1}C_P(a_P F_P + a_G F_G) - (m_P + m_V)C_P + t_{DP} m_D C_D \\
&\quad + \alpha m_A (A_P + A_G) \\
&\quad + r_P\beta[m_A n(A_P + A_G) + m_F(F_P + F_G + F_D)],
\end{aligned} \tag{6.32}
$$

$$
\begin{aligned}
dC_D/dt &= \text{(supply from primary production)} \\
&\quad - \text{(loss by uptake of generalists)} \\
&\quad - \text{(loss by uptake of DOC specialists)} \\
&\quad - \text{(loss due to unspecified processes)} \\
&\quad + \text{(conversion from POC)} \\
&\quad + \text{(regeneration during bacterial mortality)} \\
&= s_D + f_D\alpha h(A_P + A_G) - {e_D}^{-1}u_G C_D\beta F_G - {e_D}^{-1}u_D C_D\beta F_D \\
&\quad - m_D C_D + t_{PD} m_P C_P \\
&\quad + r_D\beta[m_A n(A_P + A_G) + m_F(F_P + F_G + F_D)],
\end{aligned} \tag{6.33}
$$

where s_P and s_D are constant supply rate of POC and DOC, m_P and m_D are the loss rate, but the fraction t_{PD} and t_{DP} of lost POC and DOC are converted to DOC and POC, respectively. There is also a recycle of organic carbon along bacterial mortality. The fraction r_P and r_D of lost bacterial carbon are recycled as POC and DOC, respectively.

In this model, competition between POC specialists and generalists for particle colonization, and competition between DOC specialists and generalists for DOC uptake are considered. To allow the possibility of coexistence between specialists and generalists, advantages for each type of specialists are assumed: (1) POC specialists are better competitors for colonization ($a_P > a_G$), and (2) DOC specialists are better competitors for DOC uptake ($u_D > u_G$).

There are a few key parameters in this model as the determinants of bacterial community compositions: supply rate of two resources (s_P and s_D), quality of these resources represented by growth efficiency (e_P and e_D), and the fate of the hydrolyzed carbon (f_F and f_D) that is related to the interdependency between DOC users (DOC specialists and generalists) and POC users (POC specialists and generalist). Although these parameters would dynamically change during phytoplankton bloom and then bacterial community composition shifts at non-equilibrium state, it is also possible to assume that these parameters' change is slower than the dynamics of bacterial community. Then, the shift of bacterial community composition during phytoplankton bloom can be understood as the shifts of equilibrium state of bacterial community along with gradual changes of parameter values, e.g., the increase of POC supply (s_P) along with the increase of large-size phytoplankton such as diatom. Therefore, the goal of mathematical analysis for this purpose is to check the dependency of positivity and local asymptotical stability of each equilibrium on parameter values. There are seven distinct equilibria with different bacterial composition for this system: no persistence of any of three bacterial groups ($\mathbf{E}_N$), POC specialists only ($\mathbf{E}_P$), DOC specialists only ($\mathbf{E}_D$), generalists only ($\mathbf{E}_G$), and three combinations of two group coexistence ($\mathbf{E}_{PD}$, $\mathbf{E}_{GD}$, $\mathbf{E}_{GP}$). Note that the coexistence of three different groups together is impossible because there are only two independent resources for this competitive system. Fortunately, it is possible to mathematically write down all of these seven equilibria (not shown here but see Appendix in Miki and Yamamura 2005b). For each of these equilibria, local asymptotical stability (LAS) conditions are determined as algebraic inequalities using Routh–Hurwitz criteria (see Smith and Waltman 1995). Then, which equilibrium is locally asymptotically stable, depending on parameter values, is investigated through numerically checking these inequalities. No multistable regions are observed in given ranges of parameter values so only one of the seven equilibria is realized for each parameter value set.

The simplest scenario to mimic the change of carbon supply during phytoplankton bloom is the increase of POC supply (s_P) with keeping all of the other parameters constant including DOC supply (s_D, but see the Sect. 6.3.2.1). The model analysis predicts that bacterial community shifts from (1) the dominance of DOC specialists ($\mathbf{E}_D$ is LAS, region D), (2) the emergence of generalists coexisting with DOC specialists ($\mathbf{E}_{GD}$ is LAS, region GD), (3) the disappearance of DOC specialist but the dominance of generalists ($\mathbf{E}_G$ is LAS, region G), and (4) the emergence of POC specialists ($\mathbf{E}_{GP}$ is LAS, region GP) (Fig. 6.8b). This predicted pattern fits well with the community dynamics observed during an experimentally induced phytoplankton bloom (Riemann and Winding 2001). These results imply the effectiveness of this modeling framework that distinguishes (1) free-living and particle-associated cells and (2) different strategies in terms of carbon utilization in bacterial community. However, it remains unclear how much diversity and complexity in bacterial community should be considered in modeling, which is in reality found to be very high (Venter et al. 2004; DeLong et al. 2006). This question is focused on later in the next part of this chapter (see the Sect. 6.3.2).

6.3.1.2 Mathematical Model for Interactions between Bacteria, Phage, and Protozoa

For this second model, instead of focusing on the details in the interactions between organic carbon and bacterial community, the complex interactions between bacteria, viruses as parasites on bacteria, and protozoa as predators of bacteria are considered. These three components are the basis for microbial loop and microbial food webs (Miki and Jacquet 2008). Here, a basic model, which can be easily extended, is presented here. Although population dynamics and coexistence of these three microbial players are the main focus, the model should consider mass balance since microbial population dynamics are directly linked to nutrient and carbon fluxes. Here the model is carbon-based for simplicity. The microbial food web consists of five components: organic carbon resource (C), two cell states of heterotrophic bacteria—uninfected (U) and infected (I) cells, virus (V) infecting on bacteria, and protozoa (or heterotrophic nanoflagellate HNF) (H). It is natural to use the unit of density for microbial components while using the unit of carbon concentration for C. More specifically, in order to mimic natural abundance and concentration of them, the unit for C is μgC L^{-1}, and the common unit for U, I, V, and H is 10^9 cells (or particles) L^{-1}. To meet with mass balance for carbon, the cell carbon content is parameterized as c_U, c_I, c_V, and c_H [microgC (10^9 cells)$^{-1}$] (see Eqs. 6.35, 6.37, and 6.38). By assuming that a single particle of virus attaches to an uninfected cell and converts it into an infected cell, $c_I = c_U + c_V$.

The dynamics of carbon resource is given by,

$$\begin{aligned} dC/dt &= (\text{constant supply from phytoplankton}) - (\text{loss by dilution}) \\ &\quad - (\text{loss by bacterial consumption}) \\ &\quad - (\text{recycle along with viral decay}) \\ &= I_0 - qC - r_U CU + \beta\lambda c_V V, \end{aligned} \tag{6.34}$$

where λ is the viral decay rate and β is the recycling ratio.

The dynamics of bacteria (U and I) are given by,

$$
\begin{aligned}
dU/dt &= (\text{growth by carbon consumption}) - (\text{loss by protozoa predation}) \\
&\quad - (\text{loss by new infection}) \\
&= e_B \frac{r_U C}{c_U} U - gUH - \gamma UV,
\end{aligned} \tag{6.35}
$$

$$
\begin{aligned}
dI/dt &= (\text{increase by new infection}) - (\text{loss by cell lysis}) \\
&\quad - (\text{loss by protozoa predation}) \\
&= \gamma UV - \frac{1}{\tau} I - gIH,
\end{aligned} \tag{6.36}
$$

where e_B is the bacterial growth efficiency, g and γ are predation rate and infection rate, respectively, and τ is the expected development time of virus from infection to lytic production of new viral particles by killing the host cell. Note that the inverse of τ is equal to lysis rate if τ follows the Poisson process.

The dynamics of bacterial enemies (V and H) are given by,

$$
\begin{aligned}
dV/dt &= (\text{increase by lytic production}) \\
&\quad - (\text{loss by absorption to host cells at new infection}) \\
&\quad - (\text{loss by natural decay}) \\
&= e_V \frac{c_U + c_V}{c_V} \frac{1}{\tau} I - \gamma UV - \lambda V,
\end{aligned} \tag{6.37}
$$

$$
\begin{aligned}
dH/dt &= (\text{growth by predation on bacteria}) - (\text{natural mortality}) \\
&= e_H g \frac{c_U U + (c_U + c_V) I}{c_H} H - m_H H,
\end{aligned} \tag{6.38}
$$

where e_V and e_H are growth efficiency of virus and protozoa, respectively, and m_H is the density-independent mortality of protozoa.

The motivation of modeling these interactions is to investigate the coexistence condition for viruses and protozoa; they are in fact intraguild prey (IGprey) and intraguild predator (IGpredator), respectively. This is because they exploit bacteria as the common resource (thus belongs to the identical guild) and protozoa indirectly kill viruses within infected bacterial cells. Therefore, bacteria, viruses, and protozoa form the module of incidental intraguild predation (IGP) (Miki and Yamamura 2005a; Miki and Jacquet 2008). As is known in the general theory of IGP (Holt and Polis 1997), the coexistence of viruses as IGprey and protozoa as IGpredator is realized only at the intermediated productivity (I_0) (Fig. 6.9). For further mathematical analysis, the subsystems approach is effective. We can investigate the behavior of the subsystem with bacteria and viruses only and that of the subsystem with bacteria and protozoa only. The mutual invasibility of protozoa and virus along these two subsystems represents the coexistence conditions. Interestingly, despite the predicted difficulty in the coexistence of viruses and protozoa in the IGP module, microbial ecologists had never focused on the mechanisms of their coexistence in

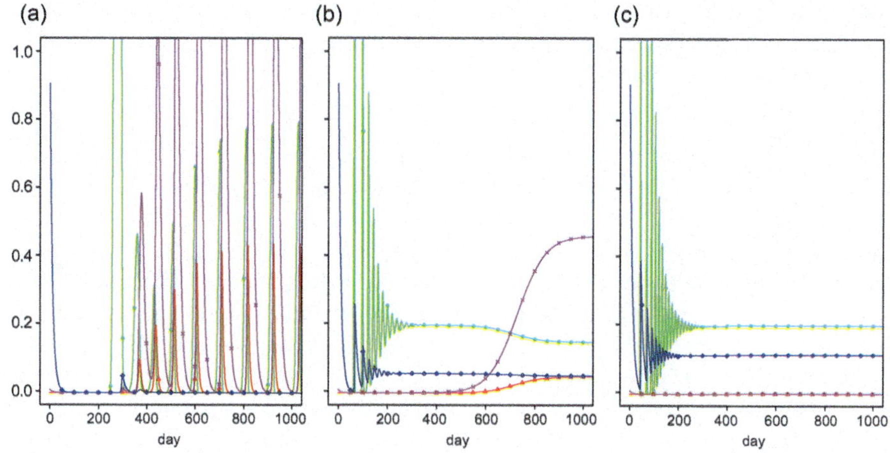

Fig. 6.9 Microbial food web model. (**a**) Protozoa go extinct with low carbon supply ($I_0 = 1.0$). (**b**) Protozoa and virus coexist with intermediate carbon supply ($I_0 = 10.0$). (**c**) Virus goes extinct with high carbon supply ($I_0 = 20.0$). Parameter values are $r_U = 0.5$, $\gamma = 0.05$, $g = 100.0$, $q = 0.01$, $\tau = 5.0$, $\lambda = 0.1$, $m_H = 0.1$, $c_U = 20.0$, $c_V = 0.2$, $c_H = 2000.0$, $I_0 = 20.0$, $\beta = 0.5$, $e_B = 0.25$, $e_V = 0.5$, and $e_H = 0.5$. U, I, V, and H are represented by green (circle), red (triangle), purple (cross), and blue (diamond) lines, with unit 10^9 cells L^{-1}, 10^9 cells L^{-1}, 10^{10} particles L^{-1}, and 10^8 cell L^{-1}, respectively

natural environments since their coexistence is quite natural along the gradient of productivity in aquatic ecosystems. So this is a paradox of IGP, implying that there are neglected mechanisms behind the coexistence of viruses and protozoa without which coexistence becomes unstable due to IGP dynamics. Potential mechanisms include the flexibility of bacterial physiology in response to productivity, which may affect infection parameters (infection rate and development rate) but it remains unexplored.

6.3.2 *Advanced Mathematical Models for Aquatic Microbial Diversity: How to Handle Microbial Diversity in Modeling?*

Accumulating evidence in the last two decades suggests the prevalence of the high taxonomic and functional diversities of microbes and the diversity and complexity of interactions between them (Venter et al. 2004; DeLong et al. 2006; Jiao and Zheng 2011). In the final section, through two examples from aquatic ecosystems, the modeling methodologies with different levels of complexity are compared. First, I compare the behavior of multiple theoretical dynamical models (TDMs) with different complexities, all of which are developed for investigating the non-equilibrium relationship between bacterial community and carbon export in the ocean during a phytoplankton bloom. Second, I also compare the behavior of a TDM with high complexity and that of a steady state model (SSM) with low complexity, both of which are intended to explain the impact of bacteriophage on bacterial carbon consumption and respiration.

6.3.2.1 How Can We Systematically Compare Multiple TDMs Without Data?

Decomposition process of POC and DOC has been parameterized as the first-order kinetics of POC or DOC concentration (i.e., "exponential decay" model, see Sarmiento and Gruber 2006) or as the rate that increases linearly with the total biomass of heterotrophic bacteria (i.e., "black box" model, Laws et al. 2000) in ecosystem models. This is a good contrast to the advanced parameterization of primary production with explicit consideration of phytoplankton diversity (Follows et al. 2007). Accumulating evidence of bacterial diversity and temporal dynamics of bacterial community composition enables us to develop more complex ecosystem models. The models with two groups, three groups, or 1000 groups show qualitatively identical behavior (Miki and Yamamura 2005b; Miki et al. 2008, 2009) but their quantitative behavior is quite different. In this section, a method for model comparison without empirical data is proposed and applied to simple marine ecosystem models.

Here, the target processes are the dynamics of bacterial community and carbon fluxes during phytoplankton bloom, which are the same as the section *Mathematical model for interactions between bacterial community, POC, and DOC*. However, not equilibrium but non-equilibrium behavior in response to phytoplankton bloom is compared between four theoretical dynamical models (TDMs) with different complexities: (1) community-based model, (2) cell-state-based model, (3) black box model, and (4) exponential decay model (Fig. 6.10). The model behavior of the community-based model is hypothetically regarded as "*reality*," and how other three models behave differently from the *reality* is investigated (Miki 2012b).

Simulating Phytoplankton Bloom As the common setting of all of the four TDMs with the dry experimental approach, temporal dynamics of primary production *PP* (*t*), POC and DOC supply during the bloom are simulated as the external forcing. The mechanistic model is used to link these three (Dunne et al. 2005; see also Miki et al. 2009),

$$s_P(PP(t)) = \frac{2f_S - f_L + f_L\sqrt{1 + 8PP(t)/PP_0}}{1 + \sqrt{1 + 8PP(t)/PP_0}} PP(t),$$
$$s_D(PP(t)) = PP(t) - s_P(PP(t)),$$

where $f_S = 0.14, f_L = 0.74$, and $PP_0 = 8.88$ [mg C m^{-3} d^{-1}].

For a simple temporal dynamics of phytoplankton bloom, *PP*(*t*) is set as

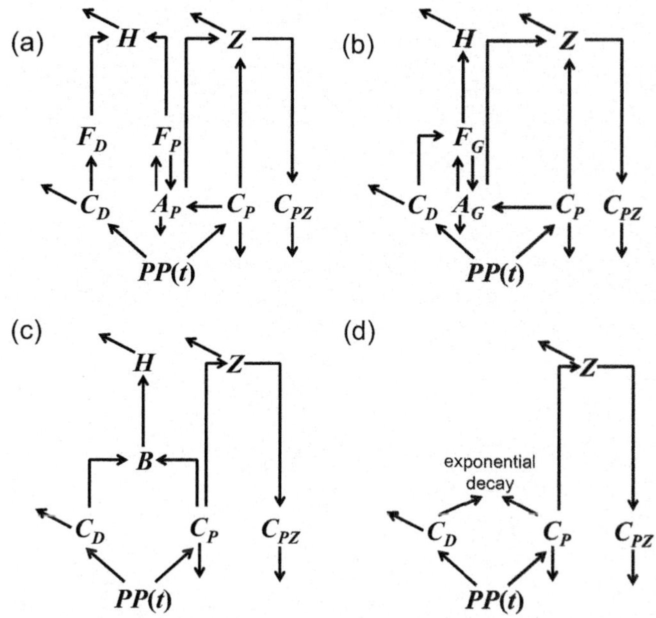

Fig. 6.10 Microbial loop models with different complexity (**a**) community-based model, (**b**) cell-state-based model, (**c**) black box model, and (**d**) exponential decay model

$$PP(t) = \begin{cases} 2.0 & \text{if } t < 0 \quad \text{before the bloom} \\ 2.0 + 1.0t & \text{if } 0 \le t \le 20 \quad \text{growing phase of the bloom} \\ 22.0 - 2.0(t - 20) & \text{if } 20 \le t \le 30 \quad \text{decaying phase of the bloom} \\ 2.0 & \text{if } t > 30 \quad \text{after the bloom} \end{cases},$$

where the unit of t and $PP(t)$ is d and mg C m^{-3} d^{-1}, respectively.

Community-Based Model Dynamics of POC and DOC supplied from phytoplankton (C_P and C_D), and POC regenerated through zooplankton grazing (C_{PZ}) are considered. C_{PZ} is assumed to be chemically and physically different from C_P so that bacteria are able to utilize C_P only. Two different strategies of bacteria as the members of bacterial community are considered, POC specialists and DOC specialists with a density of free-living cells, F_P and F_D, respectively. This is why this model is named a community-based model. The colonization of F_P to C_P forms aggregates (A_P). In addition, two different types of zooplanktons are considered: generalist bacterivorous protozoa such as heterotrophic nanoflagellates (H) on free-living bacterial cells and mesozooplankton such as copepods (Z) grazing on particles (C_P, A_P). In this model, the flexibility of bacterial functions (POC decay and DOC decay) is realized by the temporal changes in the total abundance of bacteria, the ratio of free-living and particle-associated cells, and the relative abundance of two strategies.

The dynamical equation for POC developed in the same say as the above sections,

$$\begin{aligned} dA_P/dt &= (\text{increase by bacterial colonization}) - (\text{loss by hydrolysis}) \\ &\quad - (\text{loss by vertical sinking}) - (\text{loss by zooplankton grazing}) \\ &= a_P C_P q_P^{-1} F_P - hA_P - m_V A_P - g_Z A_P Z \end{aligned} \tag{6.39}$$

where a_P and q_P are clearance rates of bacteria and carbon content per particle, respectively. Also h is the hydrolysis rate, m_V is the constant sinking rate, and g_Z is the clearance rate of mesozooplankton. It is assumed that a single cell colonizes POC, forming an aggregate, instead of the colonization of n cells (compare Eqs. 6.28 and 6.39).

Free-living cells of POC specialists and those of DOC specialists follow the following dynamics,

$$\begin{aligned} dF_P/dt &= (\text{increase along with hydrolysis}) \\ &\quad - (\text{loss by colonization to POC}) \\ &\quad - (\text{loss by zooplankton grazing}) \\ &= (1 + e_B f_F q_P q_B^{-1}) hA_P - a_P C_P q_P^{-1} F_P - g_H F_P H, \end{aligned} \tag{6.40}$$

$$\begin{aligned} dF_D/dt &= (\text{increase with DOC consumption}) \\ &\quad - (\text{loss by zooplankton grazing}) \\ &= e_B u_D C_D q_B^{-1} F_D - g_H F_D H, \end{aligned} \tag{6.41}$$

where the assumed parameters are bacterial growth efficiency (e_B), the fraction with which hydrolyzed POC is consumed by bacteria (f_F), bacterial carbon content (q_B), the clearance rate of DOC specialists (u_D), and the clearance rate of protozoa (g_H).

The dynamics of protozoa population is given by,

$$\begin{aligned} dH/dt &= (\text{increase with grazing on bacteria}) \\ &\quad - (\text{density independent and dependent mortality}) \\ &= e_H q_B q_H^{-1} g_H (F_P + F_D) H - (m_{H0} + m_H H) H, \end{aligned} \tag{6.42}$$

where the assumed parameters are growth efficiency of protozoa (e_H), carbon cell content of protozoa (q_H), density-independent mortality (m_{H0}), and density-dependent mortality($m_H H$).

Similarly, the dynamics of mesozooplankton is given by,

$$
\begin{aligned}
dZ/dt &= \text{(increase with recovery from resting eggs)} \\
&\quad + \text{(increase with grazing on particles)} \\
&\quad - \text{(density independent and dependent mortality)} \\
&= I_Z + e_Z q_Z^{-1} g_Z (C_P + (q_B + q_P) A_P) Z - (m_{Z0} + m_Z Z) Z,
\end{aligned} \tag{6.43}
$$

where e_Z, q_Z, m_{Z0}, and $m_Z Z$ are growth efficiency, carbon content per individual, density-independent rate, and density-dependent mortality, respectively. A very small rate I_Z [2.0 ind m^{-3} day^{-1}] is assumed as the supply from resting eggs for simulating the situation where mesozooplankton maintains its population with very low density with this supply before the phytoplankton bloom but its population size can increase during the bloom.

POC and DOC dynamics basically follow the same equation as Eqs. (6.32)–(6.33). In addition, regeneration of POC by mesozooplankton and that of DOC by protozoa are considered with parameters $r_{Z \to C_P}$, $r_{Z \to C_{PZ}}$, and $r_{H \to C_D}$. Also DOC is converted to non-labile carbon with a constant rate m_D due to unspecified processes (e.g., photochemical degradation). The dynamics are given by,

$$
\begin{aligned}
dC_P/dt &= \text{(supply from primary production)} \\
&\quad - \text{(loss by bacterial colonization)} - \text{(loss by sinking)} \\
&\quad - \text{(loss by grazing)} + \text{(regeneration during grazing)} \\
&= s_P(PP(t)) - a_P C_P F_P - m_V C_P - g_Z C_P Z \\
&\quad + r_{Z \to C_P} g_Z (C_P + (q_B + q_P) A_P) Z,
\end{aligned} \tag{6.44}
$$

$$
\begin{aligned}
dC_{PZ}/dt &= \text{(supply from regeneration during grazing)} \\
&\quad - \text{(loss by sinking)} \\
&= r_{Z \to C_{PZ}} g_Z (C_P + (q_B + q_P) A_P) Z - m_{PZ} Z,
\end{aligned} \tag{6.45}
$$

$$
\begin{aligned}
dC_D/dt &= \text{(supply from primary production)} \\
&\quad + \text{(release during POC hydrolysis)} \\
&\quad - \text{(loss by bacterial consumption)} \\
&\quad - \text{(loss by unspecified process)} \\
&\quad + \text{(regeneration during grazing)} \\
&= s_D(PP(t)) + (1 - f_F) q_P h A_P - u_D C_D F_D - m_D C_D \\
&\quad + r_{H \to C_D} q_B g_H (F_P + F_D) H
\end{aligned} \tag{6.46}
$$

Cell-State-Based Model This model considers a single generalist strategy only but two distinct cell states: free-living cells (F_G) and aggregates (A_G), although most of the assumptions are shared with the community-based model. The flexibility of bacterial functions in this model is realized by the changes in the total abundance of bacteria and the ratio of free-living and particle-associated cells. The dynamics are given by,

$$dA_G/dt = a_G C_P q_P^{-1} F_G - hA_G - m_V A_G - g_Z A_G Z, \tag{6.47}$$

$$dF_G/dt = \left(1 + e_B f_F q_P q_B^{-1}\right) hA_G + e_G u_G C_D q_B^{-1} F_G - a_G C_P q_P^{-1} F_G - g_H F_G H, \tag{6.48}$$

$$dH/dt = e_H q_B q_H^{-1} g_H F_G H - (m_{H0} + m_H H)H, \tag{6.49}$$

$$dZ/dt = I_Z + e_Z q_Z^{-1} g_Z (C_P + (q_B + q_P)A_G)Z - (m_{Z0} + m_Z Z)Z, \tag{6.50}$$

$$\begin{aligned} dC_P/dt = s_P(PP(t)) - a_G C_P F_G - m_V C_P - g_Z C_P Z \\ + r_{Z\to C_P} g_Z (C_P + (q_B + q_P)A_G)Z, \end{aligned} \tag{6.51}$$

$$dC_{PZ}/dt = r_{Z\to C_P} g_Z (C_P + (q_B + q_P)A_G)Z - m_{PZ} Z, \tag{6.52}$$

$$\begin{aligned} dC_D/dt = s_D(PP(t)) + (1 - f_F)q_P hA_G - u_G C_D F_G - m_D C_D \\ + r_{H\to C_D} q_B g_H F_G H. \end{aligned} \tag{6.53}$$

Black Box Model This model further simplifies the cell-state-based model by not distinguishing free-living and particle-associated cells nor POC and aggregates. Therefore, the flexibility of bacterial functions is realized by the changes in the total abundance of bacteria only. With using the clearance rate of bacteria for POC and DOC, v_B and u_B, respectively, the dynamics of bacterial abundance is given by,

$$dB/dt = e_B q_B^{-1} (f_F v_B C_P + u_B C_D)B - g_H BH. \tag{6.54}$$

The dynamics of other components can be written as,

$$dH/dt = e_H q_B q_H^{-1} g_H BH - (m_{H0} + m_H)H, \tag{6.55}$$

$$dZ/dt = I_Z + e_Z q_Z^{-1} g_Z C_P Z - (m_{Z0} + m_Z Z)Z, \tag{6.56}$$

$$dC_P/dt = s_P(PP(t)) - v_B C_P B - m_V C_P - g_Z C_P Z + r_{Z\to C_P} g_Z C_P Z, \tag{6.57}$$

$$dC_{PZ}/dt = r_{Z\to C_P} g_Z C_P Z - m_{PZ} Z, \tag{6.58}$$

$$\begin{aligned} dC_D/dt = s_D(PP(t)) + (1 - f_F)v_B C_P B - u_B C_D B - m_D C_D \\ + r_{H\to C_D} q_B g_H F_G H. \end{aligned} \tag{6.59}$$

Exponential Decay Model This model considers the population dynamics of mesozooplankton only and assumes that decay rates of POC ($k_{POC,B}$) and DOC ($k_{DOC,\,B}$) are constant, respectively. Therefore, no flexibility of bacterial functions is realized in this model. The dynamics are given by,

$$dZ/dt = I_Z + e_Z q_Z^{-1} g_Z C_P Z - (m_{Z0} + m_Z Z)Z, \tag{6.60}$$

$$dC_P/dt = s_P(PP(t)) - k_{POC,B}C_P - m_V C_P - g_Z C_P Z + r_{Z\to C_P} g_Z C_P Z, \tag{6.61}$$

$$dC_{PZ}/dt = r_{Z\to C_P} g_Z C_P Z - m_{PZ} Z, \tag{6.62}$$

$$dC_D/dt = s_D(PP(t)) + (1 - f_F)k_{POC,B}C_P - k_{DOC,B}C_D - m_D C_D. \tag{6.63}$$

The key idea for quantitatively comparing the behavior of three simpler models with those of the community-based model is to fit the initial conditions and model-specific parameter values of these three models to the equilibrium value of the community-based model with the condition before the bloom (i.e., $PP(t) = const = 2.0$). In this setting, all of these simpler models have almost the same equilibrium as the community-based model (i.e., the "reality"). Then, during the bloom, it is expected that the degree of the deviation of the model behavior from the community-based model's behavior depends on the structure of the models and the types of the flexibility of bacterial functions.

Let (A_P^*, F_P^*, F_D^*, H^*, Z^*, C_P^*, C_{PZ}^*, C_D^*) denote the equilibrium value of each component of the community-based model before the bloom ($t < 0$), which is obtained by long-term numerical integration of the ODE system (Eqs. 6.39–6.46) with the constant primary production, $PP(t) = 2.0$. For the cell-state-based model, the initial condition ($A_G(0)$, $F_G(0)$, $H(0)$, $Z(0)$, $C_P(0)$, $C_{PZ}(0)$, $C_D(0)$) is set as (A_P^*, $F_P^* + F_D^*$, H^*, Z^*, C_P^*, C_{PZ}^*, C_D^*). In order to reproduce the carbon flux from DOC to bacteria and the colonization rate to POC at $t = 0$, model-specific parameters (u_G and a_G) are determined to satisfy $u_G C_D(0) F_G(0) = u_D C_D^* F_D^*$ (Eqs.6.46 and 6.53) and $a_G C_P(0) F_G(0) = a_P C_P^* F_P^*$ (Eqs. 6.44 and 6.51). For the black box model, the initial condition ($B(0)$, $H(0)$, $Z(0)$, $C_P(0)$, $C_{PZ}(0)$, $C_D(0)$) is set as ($F_P^* + F_D^*$, H^*, Z^*, $C_P^* + A_P^*(q_P + q_B)$, C_{PZ}^*, C_D^*). The model-specific parameters (v_B and u_B) are determined by $v_B C_P(0) B(0) = hA_P^*$ (Eqs. 6.39 and 6.57) and $u_B C_D(0) B(0) = u_D C_D^* F_D^*$ (Eqs. 6.46 and 6.59), in order to reproduce the loss rate of POC and DOC by bacterial activity. Similarly, the initial condition of the exponential decay model ($Z(0)$, $C_P(0)$, $C_{PZ}(0)$, $C_D(0)$ is set as (Z^*, $C_P^* + A_P^*(q_P + q_B)$, C_{PZ}^*, C_D^*). The decay rate ($k_{POC,B}$ and $k_{DOC,B}$) are determined by $k_{POC,B} C_P(0) = hA_P^*$ (Eqs. 6.39 and 6.61) and $k_{DOC,B} C_D(0) = u_D C_D^* F_D^*$(Eqs. 6.46 and 6.63).

Although every component of the modeled systems and carbon fluxes can be compared, temporal dynamics of (1) specific rate [d^{-1}] of POC and DOC decay driven by bacteria, and (2) POC sinking rate [mgC m^{-3} d^{-1}] and DOC accumulation (DOC standing stock [mgC m^{-3}]) in $0 < t < 180$ [d] are shown here. The specific decay rates of POC and DOC change along the timeline of phytoplankton bloom, except for the exponential decay model with the constant decay rate ($k_{POC,B}$ and $k_{DOC,B}$) (Fig. 6.11a, b). The patterns of temporal changes in these decay rates differ between the four models and the deviation of these rates in three simpler models from the community-based models during $0 < t < 180$ days is large (Fig. 6.11e, f). However, counterintuitively, the variations of the model behaviors in terms of POC sinking are much smaller (Fig. 6.11c, d, g, h). This is because the total loss of POC and thus POC sinking are influenced not only by bacteria-mediated decay that shows large variations between models (Fig. 6.11e, f) but also by zooplankton grazing that

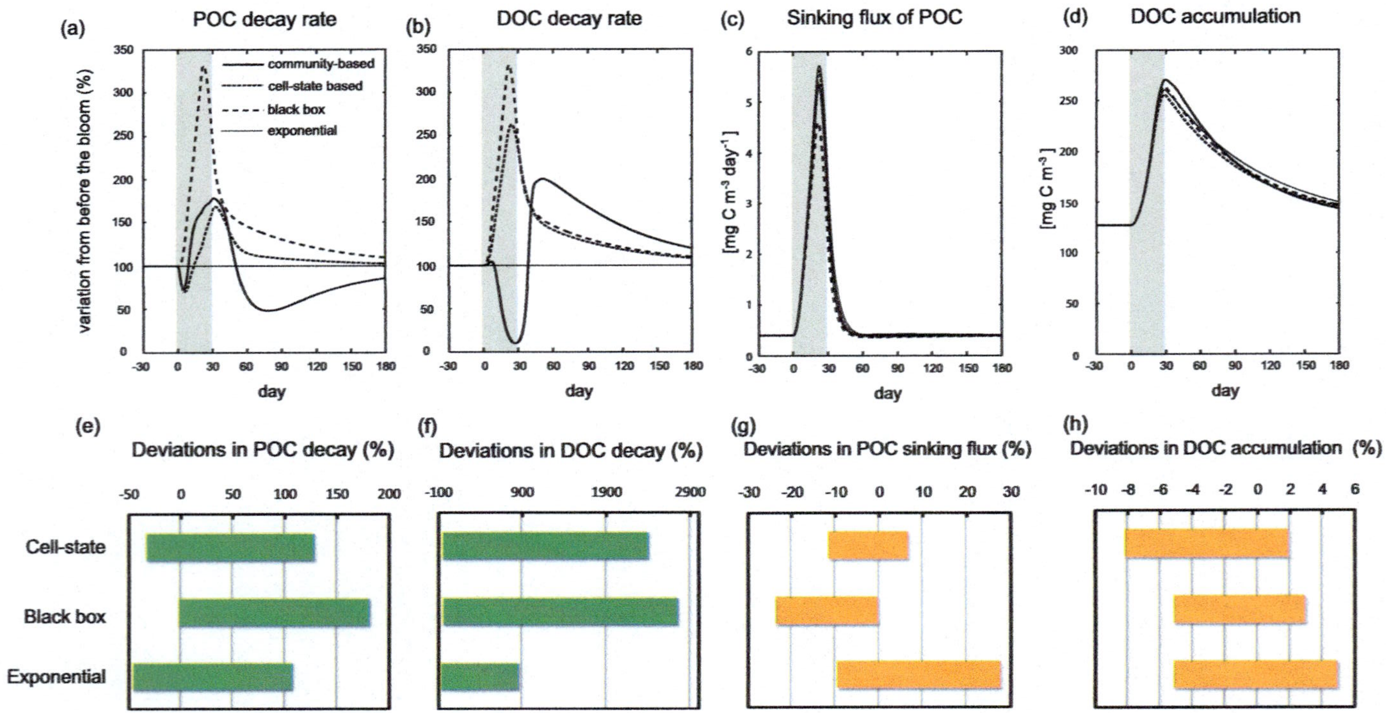

Fig. 6.11 Flexibility of bacteria functions and carbon fluxes and variations among models. **(a–d)** Variations of specific decay rates (day^{-1}) of POC and DOC, and POC sinking and DOC accumulation, after the occurrence of the bloom. **(e–h)** The ranges of deviation of the specific four indices **(a–d)** in simpler models from those in the community-based model. Parameters are $e_B = 0.25$, $e_H = e_Z = 0.5$, $f_F = 0.5$, $r_{Z \to C_P} = r_{Z \to C_{PZ}} = r_{H \to C_D} = 0.1$ (dimensionless), $m_V = 0.1$, $m_{PZ} = 0.2$, $m_{H0} = 0.1$, $m_{Z0} = 0.01$, $m_D = 0.01$ (day^{-1}), $h = 0.05$ (day^{-1}, Sarmiento and Grober 2006), $m_H = m_Z = 100$, $a_P = 0.1$, $u_D = 2.0\text{e-}4$, $g_H = 100$ (m^3 [10^9 inds.]$^{-1}$ day^{-1}), $g_Z = 9.6\text{e+}4$ (m^3 [10^9 inds.]$^{-1}$ day^{-1}, Frost 1972), $q_B = 0.015$, $q_H = 1000 q_B$ (mg C [10^9 cells]$^{-1}$), $q_P = 2000$ (mg C [10^9 cells]$^{-1}$, Frost 1972), and $q_Z = 1.0\text{e+}8$ (mg C [10^9 inds]$^{-1}$, Tande 1982)

shows fewer variations (not shown but see Miki 2012b). Note that the parameterization of zooplankton is basically common for all of the four models. Similarly, the variations in DOC accumulation are also much smaller than those in DOC decay rate. This is because the total loss of DOC is driven not only by bacteria but also by unspecified decay process (i.e., $m_D C_D$). In summary, the presence of such competitive mechanisms (e.g., zooplankton grazing and constant DOC loss) for the focal processes (POC and DOC dynamics) buffers the behavioral differences of models with different complexity of microbial components. Another lesson from these comparisons is that the temporal changes during the bloom in POC and DOC decay rates (Fig. 6.11a, b) of the community-based model is well approximated by the cell-state-based model but not by the black box model. It implies that considering the difference between cell states, that is, free-living state and particle-associated state is crucial for understanding non-equilibrium dynamics of bacterial functions in aquatic ecosystems, e.g., during phytoplankton bloom.

6.3.2.2 Is Steady State Modeling (SSM) Suitable for Understanding Non-equilibrium Dynamics in Bacteria-Mediated Carbon Cycling?

Here, instead of comparing multiple TDMs with different complexity, I will compare a TDM with a SSM. Before the comparison, I will explain the basis of SSM and also demonstrate how to develop SSM by using the example of interactions between organic carbon, bacteria, and virus. Then, I will develop a TDM for the same types of interactions but with more complexity and finally compare the behaviors of TDM with those of SSM.

Steady state modeling (SSM) has been popular in the field of ecosystem ecology (Anderson and Ducklow 2001; Grami et al. 2011), although this approach is seldom discussed in the textbook of mathematical biology. Since SSM for ecosystem relies on the mass balance at equilibrium, the mass balance is first explained here as a fundamental rule in ecosystem ecology. Let X denote an ecosystem component, either biological (e.g., bacteria) or non-biological (e.g., organic carbon), with the unit of biomass per area or volume. The temporal change of X is governed by four processes only: production, consumption, inflow, and outflow with a rate P, C, F_{in}, and F_{out}, respectively. The dynamics of X is given by,

$$dX/dt = P - C + F_{\text{in}} - F_{\text{out}}. \tag{6.64}$$

In fact, the relationship between the growth (or accumulation) rate of X (dX/dt) and P, C, F_{in}, F_{out} is called mass balance for X under mass conservation law. If the mathematical formulation is not used, the mass balance is expressed as,

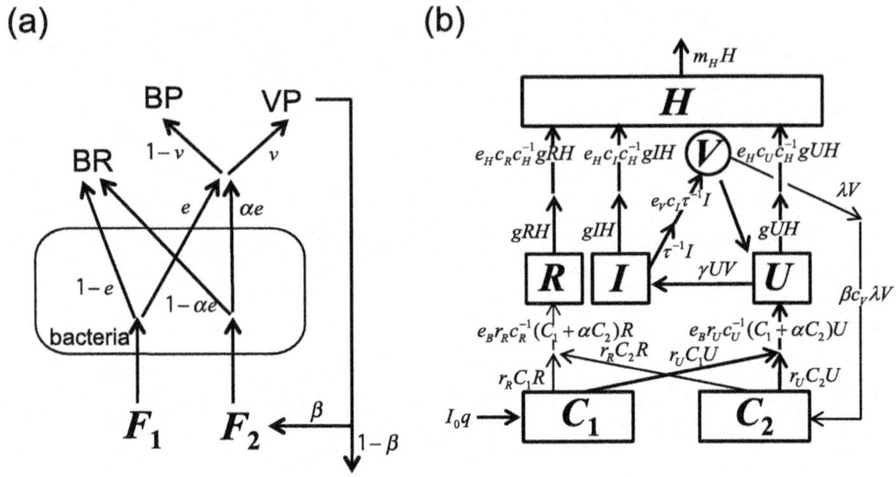

Fig. 6.12 Model diagrams for SSM and TDM. (**a**) The steady state model (SSM). (**b**) The theoretical dynamical model (TDM)

$$[\text{Accumulation rate}] = [\text{Production rate}] - [\text{Consumption rate}] + [\text{Inflow}] - [\text{Outflow}]. \tag{6.64$'$}$$

Note that the mass balance should be satisfied at every moment, even at non-equilibrium state. As a special case of the mass balance, however, the mass balance at equilibrium state (i.e., $dX/dt = 0$) can be also considered from Eqs. (6.64) and (6.64′),

$$P + F_{\text{in}} = C + F_{\text{out}}, \tag{6.65}$$

and

$$[\text{Production rate}] + [\text{Inflow rate}] = [\text{Consumption rate}] + [\text{Outflow}]. \tag{6.65$'$}$$

The Eq. (6.65) (or Eq. 6.65′) is the basis for SSM. SSM is more than the equilibrium of TDM. The distinct point of typical SSM from TDM is that SSM does not need to focus on standing biomass of ecosystem components at all but just need to focus on the fluxes between the components. This feature of SSM often results in a much simpler model in terms of the number of parameters than the corresponding TDM.

Here, a SSM that considers carbon fluxes along with the interactions between two carbon resources (C_1 and C_2), bacteria (as a black box), and virus infecting bacteria is developed (Fig. 6.12a). It is assumed that C_1 is supplied from phytoplankton and thus a good resource for bacteria while C_2 comes from regeneration from viral decay. Let F_1 and F_2 denote the bacterial consumption rate of C_1 and C_2 at equilibrium. The fraction e_B is assimilated into bacterial biomass while the other fraction $(1 - e_B)$ is respired as CO_2 when bacteria consume C_1. This assimilation fraction is assumed to be lower when bacteria consume C_2 implying its lower quality than C_2, denoted as αe_B ($\alpha < 1$). Then, the total rate of bacterial respiration (BR) is given by,

$$\text{BR} = F_1(1 - e_B) + F_2(1 - \alpha e_B). \tag{6.66}$$

The other fraction ($F_1 e_B + F_2 \alpha e_B$) is potentially a part of new bacterial biomass. However, under the steady state assumption, all of this fraction should be consumed by virus or higher trophic levels such as protozoa without being accumulated in the bacteria component. It is assumed that the fraction v is consumed by virus of which flux is defined as viral production (VP) while the other fraction $(1 - v)$ goes through higher trophic levels of which flux is defined as bacterial production (BP). VP and BP is given by,

$$\text{VP} = v(e_B F_1 + \alpha e_B F_2), \tag{6.67}$$

$$\text{BP} = (1 - v)(e_B F_1 + \alpha e_B F_2). \tag{6.68}$$

What is the fate of VP? Under the steady state assumption, VP should be balanced with loss of virus. Here it is assumed that the fraction β of the lost viral carbon is recycled as carbon resource C_2. Therefore,

$$F_2 = \beta \text{VP}. \tag{6.69}$$

These (Eqs. 6.66–6.69) are all that is needed for the SSM. Note that there are two unknown variables (F_1 and F_2).

How can I use this SSM? I can use this model to calculate an important index to characterize bacteria-mediated carbon fluxes: bacterial growth efficiency (BGE). BGE is operationally defined as the ratio of measurable BP to the total consumption of carbon (= BP + BR). That is, BGE = BP/(BP + BR). BGE is known to vary between ecosystems and ecologists are interested in the regulating mechanisms of BGE (del Giorgio and Cole 1998). For this SSM, first of all, from Eqs. (6.67) and (6.69), I can see

$$F_2 = \frac{\beta v e_B}{1 - \alpha \beta v e_B} F_1.$$

Through substituting this into Eqs. (6.66) and (6.68), BP and BR can be calculated as

$$\text{BP} = \frac{(1 - v)e_B}{1 - \alpha \beta v e_B} F_1, \text{BR} = F_1 - \text{BP}$$

Finally, BGE is given by

$$\mathrm{BGE} = \frac{\mathrm{BP}}{\mathrm{BP}+\mathrm{BR}} = \frac{(1-\nu)e_B}{1-\alpha\beta\nu e_B}. \tag{6.70}$$

Since $0 < \alpha, \beta, e_B < 1$, it is easy to see that BGE decreases with the parameter ν, which represents the strength of viral shut and this pattern is confirmed by an experimental approach (Motegi et al. 2009).

Next, a TDM that explicitly considers the population dynamics of microbial food web is developed. The model consists of seven variables: carbon biomass of resource 1 and 2 (C_1 and C_2, respectively), viral-sensitive group of bacteria with uninfected and infected cells with the density U and I, respectively, and bacterial group resistant to viral infection with the density R, virus with the density V, and protozoa with the density H. The equations for their dynamics are not shown here but every carbon flux is shown in the model diagram (Fig. 6.12b) and it is easy to formulate the ODEs. Most of the parameters have the same definitions as those in the models introduced in the second section (Eqs. 6.34–6.38).

Note that the TDM has 18 parameters while the SSM has four parameters only. For the comparison of these two models, three parameters of SSM (α, β, e_B) are shared by TDM with identical values. On the other hand, ν is a fixed parameter in the SSM but ν in the TDM is dynamically determined by the consequences of population dynamics with viral production (VP $= c_U\gamma UV$) and transfer to H ($BP = g[c_RR + c_II + c_UU]H$). Since it is not interesting to compare the equilibrium of TDM and SSM that are probably close to each other, parameter values for TDM that cause the fluctuating population dynamics are intentionally chosen (Fig. 6.13a). In order to generate 100 time series from the TDM representing the population dynamics under different environments, 100 sets of parameter values for carbon input (I_0), viral lysis rate (τ^{-1}), viral infection rate (γ), and protozoan mortality (m_H) are randomly assigned while the values of the other parameters including α, β, e_B are fixed. For each time series from TDM, the strength of viral shunt (ν) and BGE are calculated in two ways: the snapshot of ν and BGE at $t = 500$ day and long-term average $\langle\nu\rangle = \langle VP\rangle/(\langle VP\rangle + \langle BP\rangle)$ and $\langle BGE\rangle = \langle BP\rangle/(\langle BP\rangle + \langle BR\rangle)$ where $\langle\rangle$ denote the average between $t = 400$–500.

The plot of ν vs. BGE clearly demonstrates that the prediction from SSM (Eq. 6.70) well explains the inter-system variations of BGE calculated from the TDM (Fig. 6.13b). Note that the line from the SSM is not the statistical fitting to the scattered points from TDM, but is drawn with given α, β, and e_B. It is not so surprising that the SSM can explain the pattern from long-term averages of the TDM, since the long-term averages of a TDM often serve as a good approximation of the equilibrium if the system fluctuates regularly around the equilibrium. Unexpectedly, however, the SSM can even explain the pattern from the temporal snapshots of which values are extracted from non-equilibrium states of the TDM. It implies that SSM approach is potentially applicable for non-equilibrium systems in natural ecosystem, despite the strong assumption of steady state. In other words, it would act as one of the justifications for past applications of SSM to estimate carbon fluxes in natural ecosystems, which unlikely reach equilibrium (Anderson and Ducklow 2001).

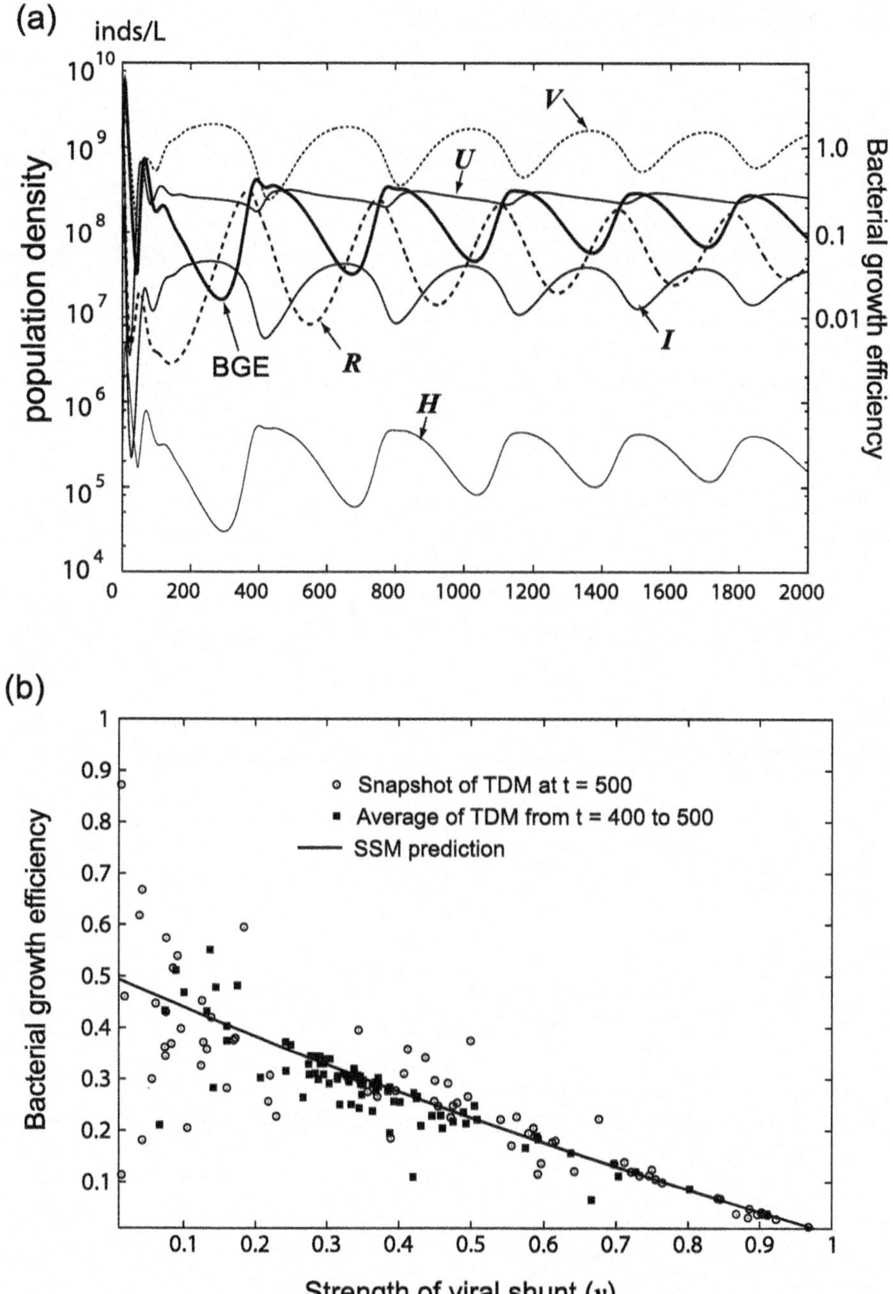

Fig. 6.13 Behaviors of SSM and TDM. (**a**) The time series from the TDM. $(r_U, r_R, \gamma, g) = (0.5, 0.2, 0.05, 100)$ [L $(10^9$ inds$)^{-1}$ day^{-1}], $(q, \tau^{-1}, \lambda, m_H) = (0.01, 0.5, 0.1, 0.1)$ [day^{-1}], $(c_U, c_I, c_R, c_V) = (20, 20.2, 20, 0.2)$[mg C $(10^9$ inds$)^{-1}$], $I_0 = 100.0$ [mgC L^{-1}], $(e, e_V, e_H, \alpha, \beta) = (0.5, 0.1, 0.3, 0.5, 0.9)$. The difference of this model from the model with Eqs. (6.34) to (6.38) is that the common loss rate (i.e., dilution rate) q is used for all components. This is the setting for chemostat. (**b**) The comparison of the behaviors of TDM and SSM

6.4 Implication and Future Directions

Now you can see multiple options for modeling the dynamics of microbes and their diversity. For the theoretical dynamical modeling (TDM), there is a freedom for the choice of structural complexity to describe microbial communities, e.g., how many functional groups and how many cell states (e.g., free-living vs particle-attached, or uninfected vs infected) are considered. When avoiding the inflation of the number of parameters unknown, the steady state modeling (SSM) is a good alternative choice, especially for understanding ecosystem level properties. One of the disadvantages of SSM would be that the degree of partitioning of one flux into multiple fluxes along trophic interactions in a food web (e.g., partitioning of bacterial carbon into BP and VP) is given as a fixed parameter (e.g., v in the above model), while this type of partitioning is the consequence of dynamics in TDM.

There are other growing approaches for handling complexity of microbial communities. Bioinformatics modeling (BIM), which uses the whole genome functional information such as richness of clusters of orthologous genes (COGs) and metabolic pathways (e.g., KEEG pathways), has been applied for predicting the relationship between bacterial diversity and ecosystem functions (Miki et al. 2014) and the interspecific interactions between microbes (Takemoto and Aie 2017). Empirical dynamic modeling (EDM) is the approach to infer causal–effect relationships from time series data, which is based on dynamical system theory but without any specific dynamical equations (Chang et al. 2017). The combination of TDM, SSM, BIM, and EDM will be a promising approach for better understand and forecast the complex behavior of microbial communities and microbe-mediated ecosystem processes.

References

Anderson TR, Ducklow HW (2001) Microbial loop carbon cycling in ocean environments studied using a simple steady-state model. Aquat Microb Ecol 26:37–49

Azam F, Fenchel T, Filed JG, Gray JS, Meyer-Reil TF (1983) The ecological role of water-column microbes in the sea. Mar Ecol Prog Ser 10:257–263

Barabas G, D'Andrea R, Stump SM (2018) Chesson's coexistence theory. Ecol Monogr 2018. https://doi.org/10.1002/ecm.1302

Bar-On YM, Phillips R, Milo R (2018) The biomass distribution on earth. PNAS. https://doi.org/10.1073/pnas.1711842115

Berendse F (1994) Litter decomposability—a neglected component of plant fitness. J Ecol 82:187–190

Bever JD (1997) Incorporating the soil community into plant population dynamics: the utility of the feedback approach. J Ecol 85:651–573

Bever JD (2003) Soil community feedback and the coexistence of competitors: conceptual frameworks and empirical test. New Phytol 157:465–473

Brinkman EP, van der Putten WH, Bakker E, Verhoeven KJF (2010) Plant-soil feedback: experimental approaches, statistical analyses and ecological interpretations. J Ecol 98:1063–1073

Chang C-W, Ushio M, C-h H (2017) Empirical dynamic modeling for beginners. Ecol Res 32:785–796

Chesson P (2000) Mechanisms of maintenance of species diversity. Annu Rev Ecol Syst 31:343–366

Clark BR, Hartley SE, Suding KN, de Mazancourt C (2005) The effect of recycling on plant competitive hierarchies. Am Nat 165:609–622

del Giorgio PA, Cole JJ (1998) Bacterial growth efficiency in natural aquatic systems. Ann. Rev. Ecol Syst 9:503–541

DeLong EF et al (2006) Community genomics among stratified microbial assemblages in the ocean's interior. Science 311. https://doi.org/10.1126/science.1120250

Ducklow HW, Steinberg DK, Buesseler KO (2001) Upper Ocean carbon export and the biological pump. Oceanography (Wash DOC) 14:50–58

Dunne JP, Armstrong RA, Gnanadesikan A, Sarmiento JL (2005) Empirical and mechanistic models for the particle export ratio. Global Biogeochem Cycles 19:GB4026

Eppinga MB, Kaproth MA, Collins AR, Molofsky J (2011) Litter feedbacks, evolutionary change and exotic plant invasion. J Ecol 99:503–514

Eppinga MB, Baudena M, Johnson DJ, Jian J, Mack KML, Strand AE, Bever JD (2018) Frequency-dependent feedback constrains plant community coexistence. Nat Ecol Evol 2:1403–1407

Fierer N, Bradford MA, Jackson RB (2009) Toward an ecological classification of soil bacteria. Ecology 88:1354–1364

Follows MJ, Dutkiewicz S, Grant S, Chisholm SW (2007) Emergent biogeography of microbial communities in a model ocean. Science 315:1843–1846

Frost BW (1972) Effects of size and concentration of food particles on the feeding behavior of the marine planktonic copepod *Calanus pacificus*. Limnol Oceanogr 17:805–815

Fuhrman JA (1999) Marine viruses and their biogeochemical and ecological effects. Nature 399:541–548

Gallien L, Landi P, Hui C, Richardson DM (2017) Emergence of weak-intransive competition through adaptive diversification and eco-evolutionary feedbacks. J Ecol. https://doi.org/10.1111/1365-2745.12961

Grami B, Rasconi S, Niquil N, Jobard M, Saint-Beat B, Sime-Ngando T (2011) Functional effects of parasite on food web properties during the spring diatom bloom in Lake Pavin: a linear inverse modeling analysis. PLoS One 6:e23273

Holt R, Polis GA (1997) A theoretical framework for intraguild predation. Am Nat 149:745–764

Horner-Devine MC, Bohannan BJM (2006) Unifying ecology to include all creatures great and small. Trends Ecol Evol 21. https://doi.org/10.1016/j.tree.2006.07.008

Hsu SB, Hubbell S, Waltman P (1977) A mathematical theory for single-nutrient competition in continuous culture of micro-organisms. SIAM J Appl Math 32:366–383

Hsu SB, Cheng KS, Hubbell SP (1981) Exploitative competition of microorganisms for two complementary nutrients in continuous cultures. SIAM J Appl Math 41:422–444

Janzen DH (1970) Herbivores and the number of tree species in tropical forest. Am Nat 104:501–528

Jiao N, Zheng Q (2011) The microbial carbon pump: from genes to ecosystems. Appl Environ Microbiol 77:7439–7444

Jones CG, Lawton JH, Shachak M (1994) Organisms as ecosystem engineers. In: Ecosystem management. Springer, Cham, pp 130–147

Kagami M, Van Donk E, de Bruin A, Rijkeboer M, Ibelings BW (2004) Daphnia can protect diatoms from fungal parasitism. Limnol Oceanogr 49:680–685

Kagami M, Miki T, Takimoto G (2014) Mycoloop: chytrids in aquatic food webs. Front Microbiol 5:166

Ke P-J, Miki T (2015) Incorporating the soil environment and microbial community into plant competition theory. Front Microbiol 6:1066

Ke P-J, Miki T, Ding D-S (2014) The soil microbial community predicts the importance of plant traits in plant-soil feedback. New Phytol 206:329–341

Klironomos JN (2002) Feedback with soil biota contributes to plant rarity and invasiveness in communities. Nature 417:67–70

Kwon EY, Primeau F, Sarmiento JL (2009) The impact of remineralization depth on the air-sea carbon balance. Nat Geosci. https://doi.org/10.1038/NGEO612
Lankau RA, Wheeler E, Bennett AE, Strauss SY (2011) Plant-soil feedbacks contribute to an intransive competitive network that promotes both genetic and species diversity. J Ecol 99:176–185
Laws EA, Falkowski PG, Simith WO, Ducklow H, McCarthy JJ (2000) Temperature effects on export production in the open ocean. Global Biogeochem Cycles 14:1231–1146
Lenski RE, Levin BR (1985) Constraints on the coevolution of bacteria and virulent phage: a model, some experiments, and predictions for natural communities. Am Nat 125:585–602
Levine JM, Bascompte J, Adler PB, Allesina S (2017) Beyond pairwise mechanisms of species coexistence in complex communities. Nature. https://doi.org/10.1038/nature22898
Lindeman RL (1942) The trophic-dynamic aspect of ecology. Ecology 23:399–416
Lovelock JE (1972) Gaia as seen through the atmosphere. Atmos Environ 6:579–580
Lovelock JE, Margulis L (1974) Atmospheric homeostasis by and for the biosphere: the Gaia hypothesis. Tellus 26:2–10
Mangan SA, Schnitzer SA, Herre EA, Mack KML, Valencia MC, Sanchez EI, Bever JD (2010) Negative plant-soil feedback predicts tree-species relative abundance in a tropical forest. Nature 466:752–755
Mariotte P, Mehrabi Z, Bezemer TM, De Deyn GB et al (2017) Plant-soil feedback: bridging natural and agricultural sciences. Trends Ecol Evol 33:129–142
Matthews B, De Meester L, Jones CG, Ibelings BW, Bouma TJ, Nuutinen V, de Koppel J, Obling-Smee J (2014) Under niche construction: an operational bridge between ecology, evolution, and ecosystem science. Ecol Monogr 84:245–263
Miki T (2012a) Microbe-mediated plant-soil feedback and its roles in a changing world. Ecol Res. https://doi.org/10.1007/s11284-012-0937-5
Miki T (2012b) Bacterial community should be considered in ecosystem model in the ocean? Comparison between community-based model, black box model, and exponential decay model. Interdiscip Stud Environ Chem 6:407–418
Miki T, Doi H (2016) Leaf phenological shifts and plant-microbe-soil interactions can determine forest productivity and nutrient cycling under climate change in an ecosystem model. Ecol Res 31. https://doi.org/10.1007/s11284-016-1333-3
Miki T, Jacquet S (2008) Complex interactions in the microbial world: underexplored key links between viruses, bacteria and protozoan grazers in aquatic environments. Aquat Microb Ecol 51:195–208
Miki T, Kondoh M (2002) Feedbacks between nutrient cycling and vegetation predict plant species coexistence and invasion. Ecol Lett 5:624–633
Miki T, Yamamura N (2005a) Intraguild predation reduces bacterial species richness and loosens the viral loop in aquatic systems: 'kill the killer of the winner' hypothesis. Aquat Microb Ecol 40:1–12
Miki T, Yamamura N (2005b) Theoretical model for interactions between particle-associated and free-living bacteria to predict functional composition and succession in the bacterial communities. Aquat Microb Ecol 39:35–46
Miki T, Yokokawa T, Nagata T, Yamamura N (2008) Immigration of prokaryotes to local environments enhances remineralization efficiency of sinking particles: a metacommunity model. Mar Ecol Progr Ser 366:1–14
Miki T, Giuggioli L, Kobayashi Y, Nagata T, Levin SA (2009) Vertically-structured prokaryotic community can control the efficiency of the biological pump in the oceans. Theor Ecol 2009:199–216
Miki T, Ushio M, Fukui S, Kondoh M (2010) Functional diversity of microbial decomposers facilitates plant coexistence in a plant-microbe-soil feedback model. Proc Natl Acad Sci U S A 107:14251–14256
Miki T, Takimoto G, Kagami M (2011) Roles of parasitic fungi in aquatic food webs: a theoretical approach. Freshw Biol 56:1173–1183

Miki T, Yokokawa T, Matsui K (2014) Biodiversity and multifunctionality in a microbial community: a novel theoretical approach to quantify functional redundancy. Proc R Soc B 281 (1776):20132498

Motegi C, Nagata T, Miki T, Weinbauer MG, Legendre L, Rassoulzadegan F (2009) Viral control of bacterial growth efficiency in marine pelagic environments. Limnol Oceanogr 54:1901–1910

Pomeroy LR (1974) The ocean's food web, a changing paradigm. Bioscience 24:499–504

Pomeroy LR, Williams PJI, Azam F, Hobbie JE (2007) The microbial loop. Oceanography 20:28–33

Riemann L, Winding A (2001) Community dynamics of free-living and particle-associated bacterial assemblages during a freshwater phytoplankton bloom. Microb Ecol 42:274–285

Saavedra S, Rohr RP, Bascompte J, Godoy O, Kraft NJB, Levine JM (2017) A structural approach for understanding multispecies coexistence. Ecol Monogr 87:470–486

Sarmiento JL, Gruber N (2006) Ocean biogeochemical dynamics. Princeton University Press, Princeton

Schnitzer SA, Klironomos JN, HilleRisLmbers J, Kinkel LL, Reich PB, Xiao K, Riling MX, Sikes BA, Callaway RM, Mangan SA et al (2011) Soil microbes drive the classic plant diversity-productivity patterns. Ecology 92:296–303

Smith HL, Waltman P (1995) The theory of the chemostat: dynamics of microbial competition. Cambridge University Press, Cambridge

Strickland MS, Lauber C, Fierer N, Bradford MA (2009) Testing the functional significance of microbial community composition. Ecology 90:441–451

Suttle CA (2007) Marine viruses—major players in global ecosystems. Nat Rev Microbiol 5:801–812

Takemoto K, Aie K (2017) Limitations of a metabolic network-based reverse ecology method for inferring host-pathogen interactions. BMC Bioinformat 18:278

Tande KS (1982) Ecological investigation on the zooplankton community of Balsfjorden, northern Norway: generation cycles, and variations in body weight and body content of carbon and nitrogen related to overwintering and reproduction in the copepod *Calanus finmarchicus*. J Exp Mar Biol Ecol 62:129–142

Tansley AG (1935) The use and abuse of vegetational concepts and terms. Ecology 16:284–307

Thingstad TF (2000) Elements of a theory for the mechanisms controlling abundance, diversity, and biogeochemical role of lytic bacterial viruses in aquatic systems. Limnol Oceanogr 45:1320–1328

Tilman D (1980) A graphical-mechanistic approach to competition and predation. Am Nat 116:362–393

Tilman D (1994) Competition and biodiversity in spatially structured habitats. Ecology 75:2–16

Toju H, Peay KG, Yamamichi M et al (2018) Core microbiomes for sustainable agroecosystems. Nat Plant 4:47–257

Ushio M, Miki T, Balser TC (2013) A coexisting fungal-bacterial community stabilizes soil decomposition activity in a microcosm experiment. PLoS One. https://doi.org/10.1371/journal.pone.0080320

van der Putten HW, van Dijk C, BAM P (1993) Plant-specific soil-borne diseases contribute to succession in foredune vegetation. Nature 362:53–56

Venter JC et al (2004) Environmental genome shotgun sequencing of the Sargasso Sea. Science 304. https://doi.org/10.1126/science.1093857

Watson AJ, Lovelock JE (1983) Biological homeostasis of the global environment: the parable of Daisyworld. Telllus 35B:284–289

Weitz JS (2016) Quantitative viral ecology: dynamics of viruses and their microbial hosts. Princeton University Press, Princeton

Yachi S, Loreau M (1999) Biodiversity and ecosystem productivity in a fluctuating environment: the insurance hypothesis. Proc Natl Acad Sci U S A 96:1463–1468

Part III
Diversity of Inter-Specific Interaction

Chapter 7
Ecology and Evolution of Density-Dependence

Kazutaka Kawatsu

Abstract In nature, ecological communities often consist of many species that are tightly coupled with others in diverse ecological relationships. This ecological complexity has attracted keen attention both in, but different between evolutionary biology and community ecology. In this chapter, I first review evolutionary view of ecological complexity: tight links with other species yield strong selection pressure, diversifying adaptations of interspecific interactions. Then, I explain how the pattern of adaptations in interspecific interaction can affect community dynamics. In particular, diversity of functional response or density-dependence in interspecific interaction strongly alters relationship between complexity and community stability, solving the problematic 'complexity-stability' debate in community ecology.

Keywords Complexity-stability debate · Density-dependence variation · Frequency-dependent selection · Functional response

7.1 Introduction

In nature, as seen in tropical rainforests and coral reefs, a number of species tightly interact with each other in various ways, forming complex ecological networks. Ecological theory predicts that ecological interactions between species (i.e. interspecific interactions) drive population and community dynamics and that the shape of interspecific interactions is the key to species coexistence (e.g. Wootton and Emmerson 2005; Allesina and Tang 2012; Mougi and Kondoh 2012). Interspecific interactions are usually categorised based on the costs and benefits to the interacting species (Begon et al. 2005). For example, interactions that are beneficial (+) to one species but harmful (–) to the other are called consumer–resource, predator–prey, and parasite–host relationships. The other categories of interspecific

K. Kawatsu (✉)
Graduate School of Life Sciences, Tohoku University, Sendai, Japan

A. Mougi (ed.), *Diversity of Functional Traits and Interactions*, Theoretical Biology,
https://doi.org/10.1007/978-981-15-7953-0_7

interaction are also named according to the cost–benefit axis, including competitive (–, –), amensal (0, –), commensal (0, +), and mutualistic (+, +) interactions. This familiar cost–benefit axis is of course important in ecological theories; however, in this review, I will focus on another important but less conspicuous axis that characterises interspecific interactions: density-dependence.

An interspecific effect, which is defined as the magnitude of the changes in the population density of a recipient species (the species that receives the interaction effect) caused by a donor species (the species that exerts the interaction effect), can be separated into two parts: a numerical response and a functional response. The former is measured as the change in the density of the recipient species caused by slight changes in the density of the donor species, whereas the latter is measured by the change in the 'per capita' effect that the donor species has on the recipient (Holling 1959, 1965; Begon et al. 2005) and is related to the density-dependence of the interspecific interaction. For example, consider the benefits of a predator–prey interaction in which the recipient and the donor species are the predator and prey species, respectively. The per capita interaction effect is specified by the amount of benefit that 'one' predator individual gains from consuming a prey species. Thus, the density-dependence of an interspecific interaction can be defined as the function form of the functional response, i.e. how the per capita effect of the interaction depends on the density of other conspecifics.

Various behavioural and physiological responses proximately cause density-dependent responses in ecological relationships. In a predator–prey relationship, for example, aggregation behaviour, cryptic or warning colouration, mimicry, switching predation, group foraging, and so on are known to exhibit dependence on the density of the interacting species (e.g. reviewed in Edmunds 1974; Dawkins and Krebs 1979; Endler 1986; Stephens and Krebs 1986; Vermeij 1987; Turchin and Kareiva 1989). Thus, density-dependence in interspecific interactions, regardless of the function form, is common in ecological communities and has attracted keen attention from evolutionary biologists and community ecologists, but their interests have differed slightly. In this chapter, I first review the evolutionary view of density-dependence by addressing what kind of selection pressure evolves and maintains which type of density-dependence in interacting species. Then, I summarise the ecological view of density-dependence by discussing how the pattern of density-dependence in an interspecific interaction can affect community dynamics and structure. Finally, I introduce the recent theory of species coexistence that integrates the evolutionary and ecological views of density-dependence, that is, the diversity of density-dependence in interspecific interactions. Extending this theory to better fit ecological reality, I demonstrate that even a small difference in density-dependence among interacting species dramatically affects the emergent properties of community dynamics, solving the problematic 'complexity–stability' debate in community ecology.

7.2 Evolutionary View of Density-Dependence

In evolutionary biology, density- or state-dependent interaction has been the focus in frequency-dependent selection. By definition, selection is frequency dependent whenever correlations between the heritable variation and the fitness variation in a behavioural or physiological trait are dependent on the changes in trait distribution that are themselves generated by such correlations (Doebeli 2011). For example, consider a species in which individuals compete for food intake and can choose which foods to eat. Rewards from a food choice depend on the choices of the other individuals, and the choice to eat with fewer competitors may confer higher fitness. If this strategy of food choice is heritable, then the choice strategy spreads among the population and, ultimately, any food choice may confer equal fitness returns. Such frequency-dependence is common in ecological relationships and has been considered an important agent in driving adaptive evolution (Doebeli 2011). I here review when and how selection becomes frequency dependent, concentrating on predator–prey relationships. Then, I discuss whether a common ecological pattern of frequency-dependence and associated density-dependence exists across diverse ecological relationships, such as competition and mutualism.

Predator–prey interactions involve various behavioural and physiological traits of both the predator and the prey species, and many adaptive explanations have been proposed in evolutionary ecology. Among them, optimal foraging theory is the most prominent model used to explain the evolution of predator traits (MacArthur and Pianka 1966; Charnov 1976; reviewed in Krebs and Kacelnik 1991; Sih and Christensen 2001). The core of the optimal foraging theory is to specify the costs and benefits involved in a predation sequence, as a predator trait is assumed to have evolved to optimise the net benefit of food intake. The aforementioned explanation of food choice is a verbal model of the optimal foraging theory. Put more precisely, the model predicts that predators should be adapted to ignore less profitable prey items when more profitable ones are present and abundant (Charnov 1976). The model has been extended to capture more realistic situations, such as a patchy distribution of prey items (Iwasa et al. 1981) and variance in prey density (Caraco et al. 1980). Such optimal food choices can be observed as individual-level behaviour across diverse taxa, including crustaceans (Elner and Hughes 1978), insects (Hayashi 1988), fish (Werner and Hall 1974), birds (Krebs et al. 1977), and mammals (Gende et al. 2001). These models consider only the dependence on prey species in predator–prey interactions as the predator's behaviour changes in response to the number of prey items. There are, of course, other extensions that address cases in which other individuals of the same predator species interfere or cooperate in foraging. In general, with interference, an increase in the number of competitors decreases the foraging efficacy of the predator, whereas with cooperation, the foraging efficacy improves with the number of predator individuals (Sutherland 1983; Yamamura and Tsuji 1987).

On the other hand, the prey can also evolve counter adaptations against the predator's foraging strategy, called anti-predator strategies. Such prey strategies have been associated with various behavioural and physiological traits, including behavioural or physiological defences, aggregation behaviours, mimicry, and cryptic or warning colouration (reviewed in Edmunds 1974; Dawkins and Krebs 1979; Endler 1986; Stephens and Krebs 1986; Vermeij 1987; Turchin and Kareiva 1989). As with predator traits, anti-predator strategies can be under frequency-dependent selection but, unlike the case with predators, most studies have considered that they depend on their own density. For example, in an unpalatable prey species, the predation avoidance of warning colouration should improve with prey density because it increases the chance that a predator will associate the warning signal with unpalatability (Endler 1991). Importantly, the type of frequency-dependence that such anti-predator adaptations undergo should depend on the other traits that correlate with them. For the above example, if the prey species were palatable and warning colouration were not an honest signal, as in mimicry, then the frequency-dependence should reverse: the avoidance efficiency decreases with an increase in prey density (Mallet and Joron 1999).

For other ecological relationships, for example, competitive and mutualistic relationships, frequency-dependent selection plays a role in shaping interactions. As mutualism is often compared with consumer–resource relationships (Holland and DeAngelis 2009, 2010) and exploitative competition is defined as indirect competition for resources between consumers (Begon et al. 2005), the above discussed, optimal foraging logic would be applicable. That is, as in predator–prey interactions, the function form taken by a frequency-dependent selection should vary with its evolutionary context. Therefore, although whether positive or negative frequency-dependence dominates in these ecological relationships still remains an open question, it should be concluded that various types of frequency- and density-dependent interactions co-occur on evolutionary and ecological scales.

7.3 Ecological Impact of Density-Dependent Interaction

Unlike that evolutionary theories have focused on state-dependence as frequency-dependent selection that plays a role in adaptive evolution, population and community ecologists have focused on density-dependence in the context of species coexistence. Thus far, numerous variations have demonstrated that the type of density-dependence in an interspecific interaction is highly influential on both community dynamics and the likelihood of species coexistence (e.g. Holling 1959; Beddington 1975; DeAngelis et al. 1975; Abrams 1987; Arditi and Ginzburg 1989; Kuno 1992). As an example, consider a simple predator–prey model:

$$\frac{dN}{dt} = (b - aP)N$$

$$\frac{dP}{dt} = (eaN - d)P, \tag{7.1}$$

where N and P specify the density of prey and predator, respectively. The parameters b, e, d, and a represent the intrinsic growth rate of prey, the conversion efficiency in consuming prey, the mortality rate of predators, and the per capita attack rate, respectively. The model has no locally stable states and shows oscillatory dynamics. However, incorporating dependency on prey density into the per capita attack rate dramatically alters the model dynamics. More specifically, an attack rate saturated with the prey density destabilises the system dynamics, whereas an attack rate with a sigmoidal curve (i.e. shift from an accelerating attack rate to one decelerating with the prey density) stabilises the dynamics. In community ecology, the former is often referred to as a type II functional response and the latter as a type III response (Holling 1959). Functional response is extendible to include dependence on predator density, as in ratio-dependent models (i.e. the functional response depends on the ratio of prey density to predator density; Beddington 1975; DeAngelis et al. 1975; Arditi and Ginzburg 1989), and model analyses have demonstrated that such extensions tend to have stabilising effects on the dynamics (McCann 2012).

In summary, the theory of functional response indicates that, when the impact of an interspecific interaction increases with the density of the recipient species (i.e. positive density-dependence, hereafter PDD), harmful interspecific effects provide a regulatory effect on more abundant species and thus stabilise interspecific relationships. On the other hand, the beneficial effects of PDD exert positive feedback by providing greater benefit to more abundant species, thus destabilising community dynamics. Interestingly, such relationships between interspecific effects and the likelihood of species coexistence reverse when the impact of the interaction decreases with the density of the recipient species (negative density-dependence, hereafter NDD). That is, harmful interactions and beneficial interactions destabilise and stabilise species coexistence, respectively. There is a noteworthy empirical study that tests these predictions. Ishii and Shimada (2012) experimentally investigated the coexistence of a one-parasitoid and two-host insect system, showing that switching parasitism, in which parasitoids adaptively attack more abundant host insects, occurs at the population level and plays a role in the coexistence of the three-species system. However, it is difficult to quantify the density-dependence of an interspecific interaction in the field; thus, further studies are still needed to determine the contribution of density-dependence to species coexistence.

Density-dependence in interspecific interactions affects not only species coexistence but may also contribute to the emergent properties of the ecological community. Several theoretical studies have focused on density-dependence or the functional response of interspecific interactions in the context of 'complexity–stability' relationships, i.e. relationships between ecological complexity (species diversity and dense interspecific interaction) and the stability of species coexistence. Since Robert May's seminal paper demonstrated that a large, complex community is unlikely to be stable against infinitesimal perturbations, the complexity–stability issue has been one of the most highly debated topics in community ecology, and great efforts have been made to identify the biological and ecological mechanisms that contribute to the stability of complex communities (reviewed in Pimm 1991; McCann 2000; Namba 2015). Among these efforts, Nunney (1980) argued that, with a type III functional response in predator–prey interactions, the magnitude of the self-regulatory terms in the stability criterion increases faster than the off-diagonal terms with the number of interspecific interactions, resulting in a positive complexity–stability relationship. However, providing some counterexamples using models of low-dimensional (4–10 species) systems, Abrams and Allison (1982) argued that a positive complexity–stability relationship with a type III curve is based on unrealistic assumptions. Kondoh (2003) demonstrated that food-web persistence improves with community complexity when predators show adaptive or flexible foraging in response to prey abundance. In addition to this adaptive foraging theory, Kondoh (2007) analysed the effects of adaptive anti-predator defence (i.e. prey species adaptively invest their resources into defence in response to the risk posed by each predator) on the complexity–stability relationship and found that an adaptive predator-specific defence enhances community-level stability. However, positive complexity–stability relationships occur only under restricted conditions (Kondoh 2007).

7.4 Variation in Density-Dependence

The evolutionary view of density-dependence predicts that, as the function form of frequency-dependent selection varies with the evolutionary context for each trait, various types of density-dependence co-occur in interspecific interactions. A series of ecological theories have suggested that density-dependence in interspecific interactions plays a role in the complexity–stability relationship. However, these models were based on an implicit assumption that a density-dependent function is fixed and uniform in the community (e.g. models assuming only a type I, II, or III functional response; Nunney 1980; Abrams and Allison 1982). Two exceptional studies suggested that the variability of density-dependence plays a role in the complexity–stability relationship (Kondoh 2003, 2007), but these considered flexibility only on the predator or prey side. Thus, the role of variation in density-dependence among interaction webs in the complexity–stability relationship remains unclear.

Recently, Kawatsu and Kondoh (2018) addressed this problem and found that variation in the density-dependence of interspecific interactions is the key to determining the complexity–stability relationship. More specifically, model analysis revealed that ecological complexity stabilises species coexistence when harmful interspecific effects (e.g. predation effects on prey and competition effects) have larger density-dependence than beneficial ones (rewards in consuming prey and mutualistic effects). They also found that this pattern occurs in broader parameter settings: regardless of the sign (i.e. positive or negative), a small difference in density-dependence is enough to generate a positive complexity–stability relationship (Kawatsu and Kondoh 2018). Hereafter, extending their model to incorporate the reality of species interactions, I follow and strengthen the arguments.

Kawatsu and Kondoh (2018) considered an N-species community, whose dynamics are driven by various types of interaction, including predator–prey, competitive, and mutualistic interactions. Any pair of species is connected to each other with probability C (0:1, connectance). The population dynamics is given as:

$$\frac{dX_i}{dt} = F_i = \left(r_i + \sum_{j \in \text{pos}} B_{ij} X_j - \left(s_i X_i + \sum_{j \in \text{neg}} D_{ij} X_j \right) \right) X_i, \tag{7.2}$$

where X_i, r_i, and s_i are the density, intrinsic growth rate, and the self-regulation intensity of species i, respectively. The terms B_{ij} and D_{ij} denote the *per capita* effects of positive and negative interaction effects that species j has on species i ($i \neq j$), respectively.

Although the previous study considered that the *per capita* effect of an interspecific interaction can be dependent on the densities of species involved in the interaction, it also assumed that density-dependence occurred independently between interacting pairs (Kawatsu and Kondoh 2018). For example, in a predator–prey interaction, the per capita benefit of predator i consuming prey j becomes $B_{ij} = b_{ij} X_i{}^\wedge\beta_{ij}$, whereas the per capita cost of prey j being consumed is $D_{ji} = d_{ji} X_j{}^\wedge\delta_{ji}$ (Kawatsu and Kondoh 2018). The parameters b_{ij} and d_{ji} are the slopes of the *per capita* interspecific effects on i and j, respectively; the parameters β_{ij} and δ_{ji} determine the functional type or density-dependence of the benefit to i from j and the cost to j from i, respectively. Thus, the *per capita* effect on the recipient species decelerates with its own density when β or $\delta < 0$ (i.e. NDD). The *per capita* effects with PDD saturate, as in a type II functional response, when $0 < \beta$ or $\delta < 1$, and they accelerate when β or $\delta \geq 1$. Here, I extend this model to include the shared density-dependence between interacting pairs: $B_{ij} = b_{ij}(X_j{}^\wedge\beta_{ji})(X_i{}^\wedge\beta_{ij})$ or $b_{ij}(X_j{}^\wedge\delta_{ji})(X_i{}^\wedge\beta_{ij})$ and $D_{ji} = d_{ji}(X_i{}^\wedge\beta_{ij})(X_j{}^\wedge\delta_{ji})$ or $d_{ji}(X_i{}^\wedge\delta_{ij})(X_j{}^\wedge\delta_{ji})$. This extension should be reasonable if, in an interspecific relationship, the donor species simultaneously interacts with conspecifics as well as with the recipient species, as in interference or cooperation. In fact, the ratio-dependent and Beddington–DeAngelis functional response can be categorised as this type of density-dependence (Beddington 1975; DeAngelis et al. 1975; Arditi and Ginzburg 1989). Thus, the extension should better reflect the reality of interspecific interactions.

Following the previous studies (May 1972; Allesina and Tang 2012; Mougi and Kondoh 2012; Kawatsu and Kondoh 2018), I used the local stability of feasible equilibrium, which is easily evaluated with the community matrix M, i.e. the linearisation of the system at the coexistence equilibrium. Given that Eq. (7.2) holds that $F_i = 0$ at equilibrium, the diagonal elements of the community matrix can be obtained as partial derivatives with respect to the density of species i:

$$
\begin{aligned}
M_{ii} &= \left.\frac{\partial F_i}{\partial X_i}\right|_{X^*} \\
&= -s_i X_i^* + \sum_{j\in +,+} \beta_{ij} b_{ij} \left(X_j^*\right)^{\beta_{ji}+1} (X_i^*)^{\beta_{ij}} + \sum_{j\in +,-} \beta_{ij} b_{ij} \left(X_j^*\right)^{\delta_{ji}+1} (X_i^*)^{\beta_{ij}} \\
&\quad - \sum_{j\in -,+} \delta_{ij} d_{ij} \left(X_j^*\right)^{\beta_{ji}+1} (X_i^*)^{\delta_{ij}} - \sum_{j\in -,-} \delta_{ij} d_{ij} \left(X_j^*\right)^{\delta_{ji}+1} (X_i^*)^{\delta_{ij}},
\end{aligned} \tag{7.3}
$$

where X_i^* denotes the equilibrium density of species i. For the off-diagonal elements, when species i is significantly affected by species j, the off-diagonal elements with negative and positive interaction effects become

$$
\begin{aligned}
M_{ij} &= \left.\frac{\partial F_i}{\partial X_j}\right|_{X^*} \\
&= (\beta_{ji}+1) b_{ij} \left(X_j^*\right)^{\beta_{ji}} (X_i^*)^{\beta_{ij}+1} \text{ or } (\delta_{ji}+1) b_{ij} \left(X_j^*\right)^{\delta_{ji}} (X_i^*)^{\beta_{ij}+1} \text{ and} \\
&\quad - (\beta_{ji}+1) d_{ij} \left(X_j^*\right)^{\beta_{ji}} (X_i^*)^{\delta_{ij}+1} \text{ or } - (\delta_{ji}+1) d_{ij} \left(X_j^*\right)^{\delta_{ji}} (X_i^*)^{\delta_{ij}+1},
\end{aligned} \tag{7.4}
$$

respectively, whereas $M_{ij} = 0$ when species j has no effects on species i.

As the system is locally stable if all eigenvalues of M have negative real parts, the stability criterion of a random community matrix with $CN >> 1$, $E(M_{ij}) = 0$ and $E(M_{ij}M_{ji}) = 0$ is analytically given by

$$
\sqrt{N \cdot \mathrm{Var}(M_{ij})} < -M_{ii} \tag{7.5}
$$

(May 1972; Allesina and Tang 2012, 2015). Following the previous study (Kawatsu and Kondoh 2018), some constraints was assumed to ensure the above conditions. That is, parameters are set as constant $s(s_i = s$, $b_{ij} = b$, $d_{ij} = d$, $\beta_{ij} = \beta$, $\delta_{ij} = \delta$, and $X_i^* = X^*)$, and positive and negative interspecific interactions are set to be of the same magnitude, on average (i.e. $d = b(X^*)^{\beta-\delta}$). As $E(M_{ij}) = 0$ and $E(M_{ij}M_{ji}) = 0$ for these constraints, the diagonal elements M_{ii} and $\mathrm{Var}(M_{ij})$ are approximated by

$$
\begin{aligned}
M_{ii} &= -sX^* + \frac{(N-1)C}{4}\left((X^*)^{\beta} + (X^*)^{\delta}\right)(\beta-\delta) b (X^*)^{\beta+1} \text{ and } \mathrm{Var}(M_{ij}) \\
&= \frac{N(N-1)C}{2N^2}\left(\left((\beta+1)(X^*)^{\beta}\right)^2 + \left((\delta+1)(X^*)^{\delta}\right)^2\right)\left(b(X^*)^{\beta+1}\right)^2,
\end{aligned} \tag{7.6}
$$

respectively. Substituting Eqs. (7.6) into (7.5) and assuming $N-1 \approx N$ for $CN >> 1$, the stability criterion for communities with the shared density-dependence variation is given by

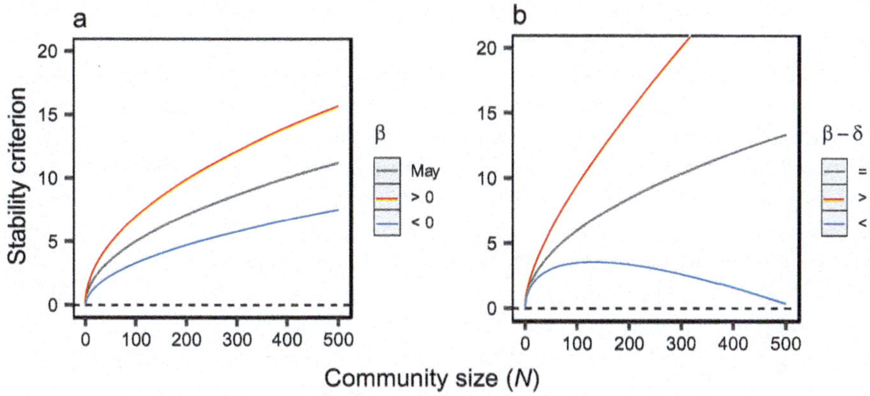

Fig. 7.1 Stability criterion for a random community with variation in shared density-dependence. The lines represent the behaviour of the right-hand side of Eq. (7.2) as a function of community size N. (**a**) Equal density-dependence between positive and negative interactions (i.e. $\beta = \delta$, $\beta = 0.0, 0.2,$ and -0.2 for grey, red, and blue lines, respectively). (**b**) The effect of density-dependence variation on the stability criterion $(\beta, \delta) = (0.1, 0.1), (0.2, 0.1)$, and $(0.1, 0.2)$ for grey, red, and blue lines, respectively. Other parameters: $s = 1.0$, $C = 1.0$, $X^* = 1.5$, and $b = 0.5$

$$s > b\sqrt{NC}$$
$$\times \left(\sqrt{\frac{\left((\beta+1)(X^*)^{\beta}\right)^2 + \left((\delta+1)(X^*)^{\delta}\right)^2}{2}} + \frac{\sqrt{NC}}{4}\left((X^*)^{\beta} + (X^*)^{\delta}\right)(\beta-\delta) \right)(X^*)^{\beta}, \tag{7.7}$$

which is identical to May's stability criterion (i.e. $s > b(NC)^{1/2}$, May 1972) when $\beta = \delta = 0$.

Figure 7.1 shows how the minimum self-regulation intensity required for the feasible coexistence to be locally stable changes with community size N (the right-hand side of Eq. 7.7). As in the previous study (Kawatsu and Kondoh 2018), the NDD condition ($\beta = \delta < 0$) simply promotes local stability, whereas the PDD condition ($\beta = \delta > 0$) decreases the stability compared with May's criterion. However, the results also show a monotonic increase regardless of the density-dependence condition, indicating that the negative relationship between complexity and stability is preserved in all cases (Fig. 7.1a). On the other hand, the stability criterion demonstrates that variation in density-dependence (i.e., $\beta - \delta \neq 0$) dramatically alters the complexity–stability relationship: when $\beta - \delta < 0$, the stability criterion shows a convex curve as a function of community size and finally falls below the stability threshold in large communities, whereas when $\beta - \delta > 0$, the criterion shows an accelerated rise (Fig. 7.1b).

The above results indicate that a positive complexity–stability relationship emerges when the density-dependence of negative interaction effects is larger than that of positive interaction effects. However, the results were limited to cases that met the conditions $E(M_{ij}) = 0$ and $E(M_{ij}M_{ji}) = 0$, where parameters are constant among species. Thus, I performed numerical simulations when relaxing the assumptions to investigate the effects of the density-dependence variation under broader conditions. The simulation was performed using the following procedure. First, the self-regulation intensity s_i and an equilibrium density of each species X_i^* were randomly chosen from uniform distributions $U(0, s_{\max})$ and $U(0, X_{\max})$, respectively. The effects of a positive interaction b_{ij} and a negative interaction d_{ij} were also chosen from uniform distributions $U(0, b_{\max})$ and $U(0, d_{\max})$, respectively. Then, based on the signs of the interaction matrix, the density-dependence parameters β_{ij} and δ_{ij} were randomly assigned from the uniform distributions $U(\beta_{\min}, \beta_{\max})$ and $U(\delta_{\min}, \delta_{\max})$, respectively. Note that the PDD and NDD conditions were defined as the mean values of the density-dependence distribution because each interspecific interaction can be PDD or NDD depending on the distribution range in the simulations. After setting parameters, the intrinsic growth rates r_i were determined to hold that $F_i = 0$. Under these relaxed assumptions, off-diagonal elements of the community matrix M can take any value without the constraints $E(M_{ij}) = 0$ and $E(M_{ij}M_{ji}) = 0$. The local stability of the community was investigated by calculating the eigenvalues of the obtained community matrix. Stability was defined as the proportion of communities in which the largest real part of the eigenvalue becomes negative among 1000 samples.

In summary, the simulation results are qualitatively correspondent to those of the previous study (Kawatsu and Kondoh 2018). Specifically, whereas a community with an equal density-dependence (i.e., $\beta = \delta$) shows a negative complexity–stability pattern similar to May's model (Fig. 7.2a), positive complexity–stability relationships arise once the density-dependence parameters meet the condition $\beta_{\text{mean}} - \delta_{\text{mean}} < 0$ (Fig. 7.2b–d). In particular, the stabilising effect was highest when the distributions of positive and negative interactions had NDD and PDD means, respectively (Fig. 7.2c), whereas it was lowest in communities in which both interactions had PDD means (Fig. 7.2d). That is, the community stability pattern was quantitatively changeable depending on the parameter setting, even when the condition was held at an equal value ($\beta_{\text{mean}} - \delta_{\text{mean}} = -0.4$ in Fig. 7.2b–d).

7.5 Conclusions

In this review, I extended the previous model to include shared density-dependence between interacting pairs as well as variations in the density-dependence of interspecific interactions. Thus, an interaction effect depends on the density of the donor species as well as that of the recipient species, similar to the ratio-dependent functional response or the Beddington–DeAngelis functional response (Beddington 1975; DeAngelis et al. 1975; Arditi and Ginzburg 1989). The results demonstrated that, despite the extension, the conclusion was coincident with that of the previous study (Kawatsu and Kondoh 2018): community stability is enhanced by PDD in

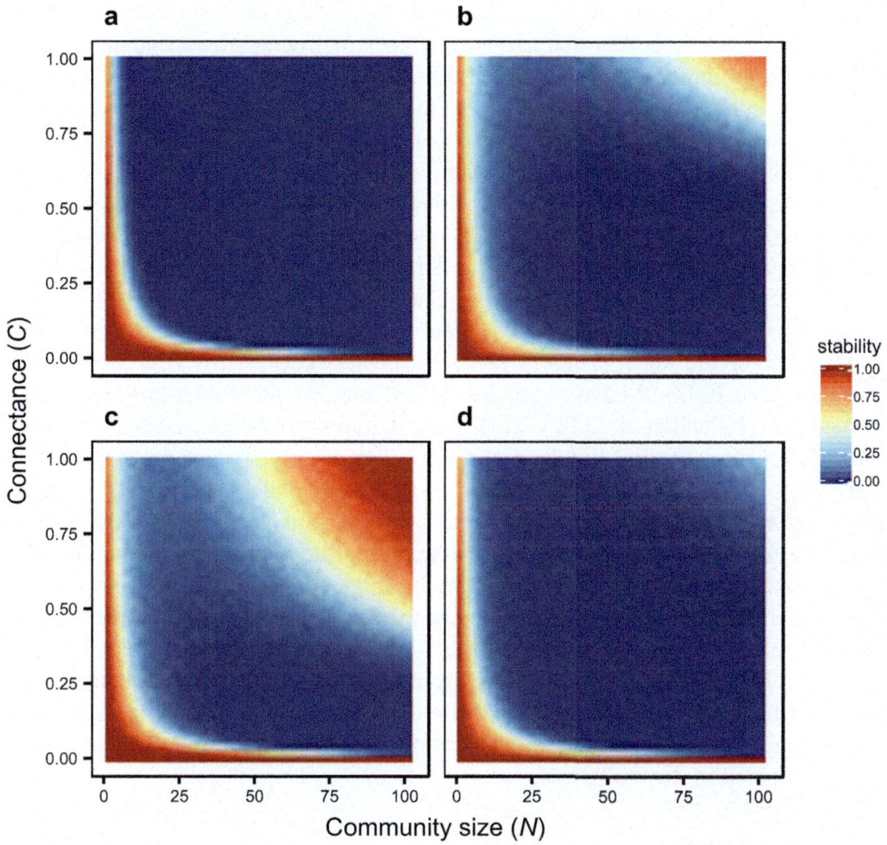

Fig. 7.2 The complexity-stability relationship in random communities with or without density-dependence variation. Different colours indicate the proportions of stable communities among 1000 samples. The distributions of density-dependence were (**a**) $\beta_{ij} = \delta_{ij} = U(0.0, 0.0)$, (**b**) $\beta_{ij} = U(-1.0, 0.0)$ and $\delta_{ij} = U(-0.3, 0.1)$, (**c**) $\beta_{ij} = U(-0.6, 0.0)$ and $\delta_{ij} = U(0.0, 0.6)$, and (**d**) $\beta_{ij} = U(-0.1, 0.3)$ and $\delta_{ij} = U(0.0, 1.0)$. Other parameters: $s_{\max} = 2.0$, $X_{\max} = 2.0$, $b_{\max} = 1.0$ and $d_{\max} = 1.0$

negative interspecific interactions and by NDD in positive interactions. Importantly, the negative complexity–stability relationship is reversed whenever the condition $\beta - \delta < 0$ is satisfied, regardless of the sign of density-dependence. That is, there is a chance for a positive complexity–stability relationship to arise when both interspecific effects are negatively or positively density dependent (Fig. 7.2b, d), because the stabilising effects of interspecific interactions (e.g. negative interactions with PDD) should cancel out the destabilising effects (e.g. positive interaction effects with PDD). As this cancelling effect accumulates with increasing ecological complexity, only a slight difference in mean density-dependence is required to yield a positive complexity–stability relationship (Fig. 7.2b–d).

In addition to this study, the previous study (Kawatsu and Kondoh 2018) confirmed that other extensions, including interaction type diversity and the classical functional responses, do not change the conclusion. Furthermore, evolutionary ecological theory indicates that as various types of frequency-dependent selection co-occur in an interspecific interaction, variation in density-dependence should be ubiquitous in an ecological community. Thus, variation in density-dependent interspecific interactions plays an important role not only under broader model spaces but also in real ecological communities. However, we know little about the patterns of density-dependence in interspecific interactions realised in nature. For example, although much effort has been made to describe the pattern of the functional response of predator attack rate (i.e. the dependency of negative interaction effects on prey density; e.g. Gross et al. 1993; Chase et al. 2000; Turchin 2003; Jeschke et al. 2004; Hixon and Jones 2005), few studies have investigated the density-dependence of the benefits that a predator species receives from its prey (Norris and Johnstone 1998). We have even less information for other relationships, such as mutualism (Holland et al. 2002). This research gap is partly due to the difficulty in quantifying interspecific effects in the field. Fortunately, recent developments in analytical techniques such as empirical dynamic modelling (EDM) approach (reviewed in Chang et al. 2017; Tsonis et al. 2018; Ushio and Kawatsu 2020) should overcome this difficulty. For example, in applying the multivariate S-map method, which enables the measurement of changing interaction effects by using only the observed time series (Deyle et al. 2016; Ushio et al. 2018; Kawatsu and Kishi 2018), we may successfully identify patterns of density-dependence that are realised in actual ecological communities. Thus, by combining these developing analytical tools, further studies investigating density-dependence in interspecific interactions will advance our understanding of how complex ecological communities are formed and maintained in the real world.

References

Abrams PA (1987) On classifying interactions between populations. Oecologia 73:272–281

Abrams PA, Allison TD (1982) Complexity, stability, and functional response. Am Nat 119:240–249

Allesina S, Tang S (2012) Stability criteria for complex ecosystems. Nature 483:205–208

Allesina S, Tang S (2015) The stability-complexity relationship at age 40: a random matrix perspective. Popul Ecol 57:63–75

Arditi R, Ginzburg LR (1989) Coupling in predator–prey dynamics: ratio-dependence. J Theor Biol 139:311–326

Beddington JR (1975) Mutual interference between parasites or predators and its effect on searching efficiency. J Anim Ecol 44:331–340

Begon M, Harper JL, Townsend CR (2005) Ecology: from individual to ecosystem, 4th edn. Blackwell Science, Oxford

Caraco T, Martindale S, Whitham TS (1980) An empirical demonstration of risk-sensitive foraging preferences. Anim Behav 28:820–830

Chang C-W, Ushio M, Hsieh C-H (2017) Empirical dynamic modelling for beginners. Ecol Res 32:785–796

Charnov EL (1976) Optimal foraging: attack strategy of a mantid. Am Nat 110:141–151

Chase JM, Abrams PA, Grover JP, Diehl S, Chesson P, Holt RD, Richards SA, Nisbet RM, Case TJ (2000) The interaction between predation and competition: a review and synthesis. Ecol Lett 5:302–315

Dawkins R, Krebs JR (1979) Arms races between and within species. Proc R Soc Lond B 205:489–511

DeAngelis DL, Mulholland PJ, O'Neil RV (1975) A model for trophic interaction. Ecology 56:881–892

Deyle ER, May RM, Munch SB, Sugihara G (2016) Tracking and forecasting ecosystem interactions in real time. Proc R Soc B 283:2258

Doebeli M (2011) Adaptive diversification. Princeton University Press, Princeton

Edmunds M (1974) Defence in animals: a survey of anti-predator defences. Longmans, London

Elner RW, Hughes RN (1978) Energy maximisation in the diet of the shore crab, *Carcinus maenas*. J Anim Ecol 47:103–116

Endler JA (1986) Defence against predators. In: Feder ME, Lauder GV (eds) Predator–prey relationships: perspectives and approaches from the study of lower vertebrates. University of Chicago Press, Chicago, pp 109–134

Endler JA (1991) Interactions between predators and prey. In: Krebs JA, Davies NB (eds) Behavioural ecology: an evolutionary approach. Blackwell Science, Oxford, pp 169–196

Gende SM, Quinn TP, Willson MF (2001) Consumption choice by bears feeding on salmon. Oecologia 127:372–382

Gross JE, Shipley LA, Hobbs NT, Spalinger DE, Wunder BA (1993) Functional response of herbivores in food-concentrated patches: tests of a mechanistic model. Ecology 74:778–791

Hayashi F (1988) Life history variation in a dobsonfly, *Protohermes grandis* (Megaloptra: Corydalidae): effects of prey availability and temperature. Freshw Biol 19:205–216

Hixon MA, Jones GP (2005) Competition, predation, and density-dependent mortality in demersal marine fishes. Ecology 86:2847–2859

Holland JN, DeAngelis DL (2009) Consumer-resource theory predicts dynamic transitions between outcomes of interspecific interactions. Ecol Lett 12:1357–1366

Holland JN, DeAngelis DL (2010) A consumer-resource approach to the density-dependent population dynamics of mutualism. Ecology 91:1286–1295

Holland JN, DeAngelist DL, Bronstein JL (2002) Population dynamics and mutualism: functional responses of benefits and costs. Am Nat 159:231–244

Holling CS (1959) Some characteristics of simple types of predation and parasitism. Can Entomol 91:385–398

Holling CS (1965) The functional response of predators to prey density and its role in mimicry and population regulation. Mem Entomol Soc Can 45:43–50

Ishii Y, Shimada M (2012) Learning predator promotes coexistence of prey species in host–parasitoid systems. Proc Natl Acad Sci USA 109:5116–5120

Iwasa Y, Hiagshi M, Yamamura N (1981) Prey distribution as a factor determining the choice of optimal foraging strategy. Am Nat 117:710–723

Jeschke JM, Kopp M, Tollrial R (2004) Consumer food systems: why type I functional responses are exclusive to filter feeders. Biol Rev 79:337–349

Kawatsu K, Kishi S (2018) Identifying critical interactions in complex dynamics between bean beetles. Oikos 4:553–560

Kawatsu K, Kondoh M (2018) Density-dependent interspecific interactions and the complexity-stability relationship. Proc R Soc B 285:20180698

Kondoh M (2003) Foraging adaptation and the relationship between food-web complexity and stability. Science 299:1388–1391

Kondoh M (2007) Anti-predator defence and the complexity-stability relationship of food webs. Proc R Soc B 274:1617–1624

Krebs JR, Kacelnik A (1991) Decision-making. In: Krebs JA, Davies NB (eds) Behavioural ecology: an evolutionary approach. Blackwell Science, Oxford, pp 105–136

Krebs JR, Erichsen JT, Webber MI, Charnov EL (1977) Optimal prey selection in the great tit (*Prus major*). Anim Behav 25:30–38

Kuno E (1992) Competitive exclusion through reproductive interference. Res Popul Ecol 34:275–284

MacArthur RH, Pianka ER (1966) On the optimal use of patchy environment. Am Nat 100:603–609

Mallet J, Joron M (1999) Evolution of diversity in warning color and mimicry: polymorphisms, shifting balance, and speciation. Annu Rev Ecol Syst 30:201–233

May RM (1972) Will a large complex system be stable? Nature 278:413–414

McCann KS (2000) The stability-diversity debate. Nature 405:228–233

McCann KS (2012) Food webs. Princeton University Press, Princeton

Mougi A, Kondoh M (2012) Diversity of interaction types and ecological community stability. Science 337:349–351

Namba T (2015) Multi-faceted approaches toward unravelling complex ecological networks. Popul Ecol 57:3–19

Norris K, Johnstone L (1998) Interference competition and the functional response of oyster-catchers searching for cockles by touch. Anim Behav 56:630–650

Nunney L (1980) The stability of complex model ecosystems. Am Nat 115:639–649

Pimm SL (1991) The balance of nature? Ecological issues in the conservation of species of communities. The University of Chicago Press, Chicago

Sih A, Christensen B (2001) Optimal diet theory: when does it work, and when and why does it fail? Anim Behav 61:379–390

Stephens DW, Krebs JR (1986) Foraging theory. Princeton University Press, Princeton

Sutherland WJ (1983) Aggregation and the 'ideal fee' distribution. J Anim Ecol 52:821–828

Tsonis AA, Deyle ER, Ye H, Sugihara G (2018) Convergent cross mapping: theory and example. In: Tsonis AA (ed) Advances in nonlinear geosciences. Springer, New York, pp 587–600

Turchin P (2003) Complex population dynamics: a theoretical/empirical synthesis. Princeton University Press, Princeton

Turchin P, Kareiva P (1989) Aggregation in *Aphis varians*: an effective strategy for reducing predation risk. Ecology 70:1008–1016

Ushio M, Kawatsu K (2020) Forecasting ecological time series using empirical modeling. In: Mougi A (ed) Diversity of functional traits and interactions-perspectives on community dynamics. Theoretical biology series. Springer, New York

Ushio M, Hsieh C-H, Masuda R, Deyle ER, Ye H, Chang C-W, Sugihara G, Kondoh M (2018) Fluctuating interaction network and time-varying stability of a natural fish community. Nature 554:360–363

Vermeij GI (1987) Evolution and escalation: an ecological history of life. Princeton University Press, Princeton

Werner EE, Hall DJ (1974) Optimal foraging and the size selection of prey by the bluegill sunfish (*Lepomis macrochirus*). Ecology 55:1216–1232

Wootton JT, Emmerson M (2005) Measurement of interaction strength in nature. Annu Rev Ecol Evol Syst 36:419–444

Yamamura N, Tsuji N (1987) Optimal patch time under exploitative competition. Am Nat 129:553–567

Chapter 8
Interaction-Type Diversity and Community Stability

Akihiko Mougi

Abstract Ecological theory predicts that a complex community formed by a number of species is inherently unstable, guiding ecologists to identify what maintains species diversity in nature. Earlier studies often assumed a community with only one interaction type, either an antagonistic, competitive, or mutualistic interaction, leaving open the question of what the diversity of interaction types contributes to the community maintenance. In this chapter, I argue that the multiple interaction types might hold the key to understanding community dynamics. Furthermore, I propose a "hybrid" community where increasing complexity leads to increased stability. I hypothesize that the diversity of species and interaction types may be the essential element of biodiversity that maintains ecological communities.

Keywords Interaction-types · Hybrid-community · Complexity · Stabilit

Ecological communities in natural ecosystems are composed of diverse species and a range of interactions. Elton (1958) argued that such natural, complex ecosystems are stable, unlike the simple ecosystems that can result from agriculture or degradation. This positive relationship between complexity and stability was accepted at that time (Odum 1953; MacArthur 1955). However, this classical view was challenged by the theoretical work of May (1972), who predicted that a more complex community network that is characterized by higher species richness or more interspecific interactions is less likely to be stable using a mathematical model of randomly constructed communities with randomly assigned interaction strengths and signs. Therefore, the contradiction between this rigorous theory and the classical view of undoubtedly complex natural ecosystems needs to be explained.

One way in which this apparent paradox could be resolved would be violation of the simplicity that was inherent in May's theory. Since the model assumed a random

---

A. Mougi
Faculty of Life and Environmental Sciences, Shimane University, Matsue, Shimane, Japan

A. Mougi (ed.), *Diversity of Functional Traits and Interactions*, Theoretical Biology,
https://doi.org/10.1007/978-981-15-7953-0_8

structure of the community network, any nonrandomness would be an easily conceivable candidate (Lawlor 1978). A number of structural, nonrandom community networks that can help species to persist in a complex community have been identified (McCann 2000; Neutel et al. 2002; Emmerson and Raffaelli 2004; Kondoh 2008; Thébault and Fontaine 2010; Stouffer and Bascompte 2011), most of which are associated with nonrandomness in the interaction network topology and interaction strength distribution. However, until recently, another nonrandom element of natural ecosystems had been overlooked: the interaction types. Many early studies considered ecological networks by exploring food webs composed of prey–predator interactions, while more recent studies have considered mutualistic networks such as plant–pollinator systems (Bascompte et al. 2003). However, while these independently developed, simplified, single-interaction-type approaches have undoubtedly contributed to our understanding of how interspecific interactions are related to population dynamics, what determines community structure, and how species diversity is maintained in nature, they have not addressed the question of what effects the multiple interaction types that coexist in nature have on natural communities.

Recent studies have demonstrated that ecological communities with different interaction types have different network structures (Thébault and Fontaine 2010) and that these different interaction types inherently play different roles in the stability of large ecological networks (Allesina and Tang 2012). More recently, theoretical and empirical studies have suggested that merging different interaction types can change the existing network structures (Lafferty et al. 2008; Melián et al. 2009; Pocock et al. 2012; Kéfi et al. 2012; Sauve et al. 2014; Toju et al. 2014). So how do these different interaction types with their different functionings interact within the same community network and affect the stability of the population dynamics? Here, I review recent research that explored the role of interaction-type diversity in community dynamics stability and argue that a diversity of interaction types can create a positive complexity–stability relationship and play a critical role in maintaining complex ecosystems—the Interaction-Type Diversity Hypothesis (ITDH) (Kondoh and Mougi 2015).

8.1 A Hybrid Module of Diverse Interaction Types

I begin by using a simple module that includes different interaction types to explore the role that the diversity of interaction types plays in ecosystem stability. While Elton used a simple prey–predator module to argue that simple ecosystems have inherent instability (Elton 1958), it remains unclear whether this argument holds in the presence of multiple interaction types.

As a classic example, consider a system in which a keystone predator regulates a competitively superior prey species. In this hybrid module, there are two different types of interspecific interactions: antagonism and competition (AC module in Fig. 8.1). In such a system, competitive interactions can have a destabilizing effect, particularly when interspecific competition is greater than intraspecific competition

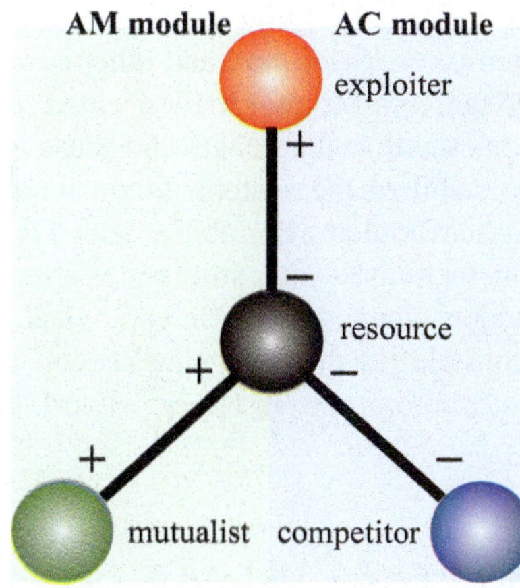

Fig. 8.1 The hybrid community module. The signs indicate antagonistic (+,−), mutualistic (+,+), and competitive (−,−) interaction types. The whole community is comprised of two qualitatively different modules: the AM module, which consists of antagonistic and mutualistic interactions, and the AC module, which consists of antagonistic and competitive interactions

(Vance 1985), and this effect can be so powerful that a globally stable predator–prey system can be destabilized and show cyclic population oscillations.

Another example is a predator–prey system that is coupled with a mutualistic interaction (Mougi and Kondoh 2014a), as occurs in terrestrial plant–herbivore–pollinator systems. This is also a hybrid module with two different types of interspecific interactions: antagonism and mutualism (AM module in Fig. 8.1). In this system, the mutualistic interactions can have a destabilizing effect, particularly when the antagonistic and mutualistic interactions are stronger, and this effect can also be so powerful that globally stable subsystems (i.e., herbivore–plant and plant–pollinator systems) can be destabilized and show cyclic population oscillations.

These instabilities are commonly explained by the indirect effect of adding further prey species to the original predator–prey system. In the former system, competitively inferior prey species can indirectly reduce the abundance of the predator by decreasing the abundance of the competitively superior prey species—and when interspecific competition is strong, this indirect effect can become so strong that the predator is greatly reduced, which can be a driver of population cycles. By contrast, in the latter system, the pollinator can indirectly increase the abundance of the herbivore by increasing the abundance of the plant, and when the mutualistic effect and exploitation are strong, the indirect effect can become so strong that the abundance of the herbivore is greatly increased, which can cause population cycles to occur.

As shown in May's theorem (May 1972), interaction strength is a key element that determines ecological stability regardless of ecosystem size. From the above examples, it can be seen that stronger interactions tend to destabilize systems, which in turn suggests that weak interactions are necessary for stabilizing systems (McCann 2000). Indeed, in the two hybrid modules described above, the systems are stable if the interactions are weak. So what happens in a hybrid module that includes all three major interaction types: antagonism, mutualism, and competition?

Let us consider a hybrid system that combines the above two submodules (Fig. 8.1), as might occur in terrestrial ecosystems or the agricultural systems considered by Elton (1958). In such a system, a plant species will interact with an herbivore, pollinator, and other competing plants. The analysis of such a system by Mitani and Mougi (2017) yielded several trivial findings: stronger interactions tend to destabilize the system and cause population cycles, while weaker interactions tend to stabilize the system. Interestingly, however, it was also found that the hybrid system could be destabilized even if each isolated submodule was stabilized by weak interactions, which strongly suggests that both hybrid systems with multiple interaction types and simple ecological communities with fewer species are inherently unstable. In the following section, I extend this thinking by considering multiple interaction types in larger network systems.

8.2 Large Hybrid Ecological Networks with Diverse Interaction Types

The question of how the diversity of interaction types affects the stability of larger ecological communities had not been addressed until relatively recently, particularly in larger systems. Interestingly, the pioneering work on larger communities with multiple interaction types was May's seminal paper (May 1972), in which the signs of interacting pairs were randomly determined in a random network to give various types of interactions, such as antagonism (+,−), mutualism (+,+), and competition (−,−). However, the aim of May's paper was not to reveal the effects of multiple interaction types on community stability. Further research by May suggested that mutualism and competition are more conducive to community stability than antagonism (May 1973). However, it is no exaggeration to say that over the next 40 years, the analysis of interaction-type diversity disappeared in preference to food-web studies in the context of the diversity–stability debate.

In an extension of May's seminal work (May 1972; Allesina and Tang 2012), the roles of major interaction types in ecological stability were examined by separating each interaction type from the random network and systematically comparing the stability of each community with only a single interaction type (Fig. 8.2). In this way, they showed that food webs are more stable and mutualistic webs are less stable, as suggested by May (1973). However, although this study showed that interaction types may affect community stability and its relationship with complexity (Allesina and Tang 2012), it remained unclear how changes in the relative frequencies or the diversity of interaction types affects community stability and the complexity–stability relationship.

While undertaking postdoctoral research at the Ryukoku University in Kyoto, Japan, Michio Kondoh and I sought to explore the role of interaction-type diversity in ecological community stability. To do this, we performed theoretical experiments in which mutualistic interactions were added to food webs in a systematic manner (Mougi and Kondoh 2012). Consider a food web composed of N species in which a proportion of all possible species pairs are interacting (represented by the

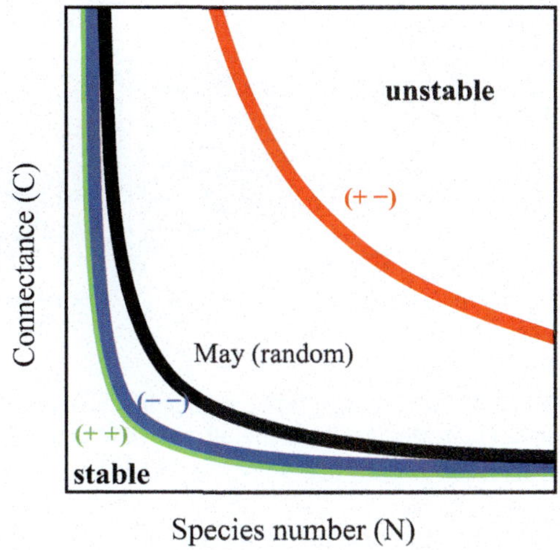

Fig. 8.2 Stability criteria for large ecological systems with different interaction types. The system is stable in the lower left and unstable in the upper right. The black line indicates May's criteria (May 1972). Red, green, and blue represent food, mutualistic, and competition webs, respectively. Modified from Allesina and Tang (2012). © 2012. Nature Publishing Group

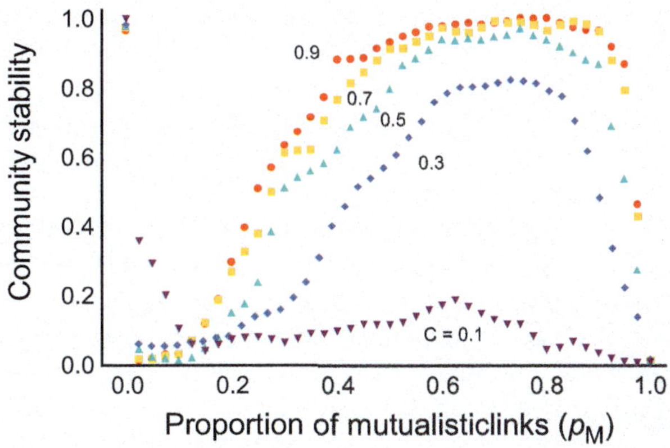

Fig. 8.3 Effects of mixing antagonistic and mutualistic interactions on community stability. Different colors indicate different levels of connectance. Modified from Mougi and Kondoh (2012. © 2012. The American Association for the Advancement of Science

connectance, *C*). Congruent with earlier theoretical studies (Allesina and Tang 2012; May 1971), we found that species might show a stable coexistence across wide ranges of *C* or *N* (Fig. 8.3). However, this was no longer the case when a small number of mutualistic interactions were added to the food web, with virtually no community being stable under this condition (Fig. 8.3). Therefore, even slight "contamination" with mutualistic interactions can completely destabilize an otherwise stable predator–prey community, clearly demonstrating the potential major effects of interaction type mixing on population dynamics. However, a further increase in the proportion of randomly chosen antagonistic links that are changed to mutualistic links (p_M) has another intriguing consequence, with the population becoming more stable and reaching its peak stability when there is a moderate mixture of both interaction types (Fig. 8.3). This suggests that a hybrid community that has both antagonism and mutualism can be more stable than communities that are biased toward either one of these interaction types.

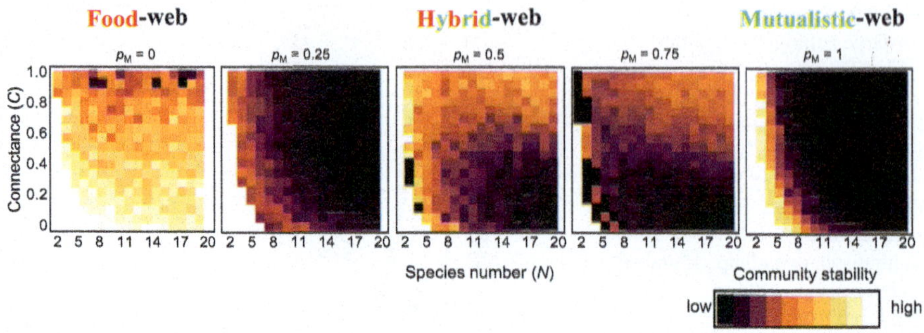

Fig. 8.4 Relationships between complexity and stability with varying proportions of mutualistic links (p_M). Modified from Mougi and Kondoh (2012). © 2012. The American Association for the Advancement of Science

A hybrid community model provides a different perspective on the complexity–stability debate (May 1972; Allesina and Tang 2015). Increased complexity (large values of *N* and *C*) destabilizes or has inconsistent effects on population stability in a nonhybrid community composed of either mutualistic or antagonistic interactions (Fig. 8.4). However, positive relationships are consistently observed in hybrid communities with a moderate mixture of these interaction types, suggesting that multispecies coexistence in a hybrid community can be maintained, rather than destabilized, by community complexity.

These intriguing effects of the hybridity of multiple interaction types are also found when another major interaction type, competition, is added to the community. All hybrid communities with a moderate mixing of two of the three major interaction types (antagonism, mutualism, and competition) can have higher levels of stability and exhibit positive complexity–stability relationships (Mougi and Kondoh 2014b). Furthermore, the moderate mixing of all three interaction types also results in a positive complexity–stability relationship (Mougi and Kondoh 2014b). However, the pattern of this relationship is qualitatively different depending on the number of interaction types and the network structure. In hybrid communities with two interaction types, the parameter region for stable coexistence becomes broader in random networks but narrower in cascade networks as complexity increases (i.e., with higher values of *N* and *C*). By contrast, in hybrid communities with three interaction types, the stable region narrows in random networks but widens in cascade networks as the complexity increases. This suggests that network structure and interaction-type diversity have synergistic effects on community stability and that interaction-type diversity may act as a stabilizer in more realistic network structures.

8.2.1 What "Kind" of Interaction Diversity Is Necessary for Community Stability?

When interpreting the ecological implications of the ITDH (Mougi and Kondoh 2012, 2014b), it is important to note that there are at least two alternative ways to define interaction type, one of which is based on the sign combination of the interspecific effect (sign-based definition), while the other is based on fitness components that are influenced by the interaction (fitness-component-based definition) (Kondoh and Mougi 2017). The former is the more traditional definition and was often used in earlier community ecological models (Allesina and Tang 2012; May 1973), whereby antagonism, mutualism, and competition were defined as interactions in which two species have (+,−), (+,+), and (−,−) effects on the counterpart's fitness, as mentioned previously. In this sign-based definition, the actual cause of the fitness change is unimportant. By contrast, the fitness-component-based definition focuses on the cause of the interspecific effect. For example, although the interactions between a plant species and its mutualistic partners, a pollinator and seed disperser, increase the fitness of all parties, the mutualistic partners are qualitatively "different" in terms of the components of plant fitness they contribute to, with the former contributing to pollination and the latter contributing to seed dispersal. Hence, having more interactions with one group (i.e., the pollinator or seed disperser) is unlikely to compensate for a shortage in the other interaction, so there should be upper limits to the fitness gain a species can receive from a single interaction type.

So which "kind" of interaction diversity is necessary for community stability under ITDH—sign, fitness component, or both? To answer this question, Kondoh and Mougi (2017) compared four types of community models with two different "kinds" of interaction types using different definitions. For simplicity, consider two types of interactions: antagonism and mutualism. In the first community model (**I**), there are antagonistic and mutualistic interactions, each of which contributes to different fitness components (both the sign- and fitness-component-based definitions). This corresponds to the model presented by Mougi and Kondoh (2012), which assumed not only that different interaction types have different sign combinations for the interaction coefficients but also that the total fitness gain a species can receive from an interaction type (antagonism or mutualism) has an upper limit and that the fitness gain a species can receive from a single antagonistic or mutualistic interaction is inversely proportional to the number of antagonistic or mutualistic partners. In the second community model (**II**), both antagonistic and mutualistic interactions contribute to the same fitness components (sign-based definition). This corresponds to the model presented by Suweis et al. (2014), which assumed that antagonism and mutualism are different only in their sign combinations and that the total fitness gain a species can receive from the interaction types has an upper limit but that the fitness gain is inversely proportional to the total number of interaction partners. In the third community model (**III**), two different kinds of antagonism contribute to different fitness components (fitness-component-based definition). Finally, in the fourth community model (**IV**), two different kinds of mutualism contribute to different fitness components (fitness-component-based definition). By comparing the effects of interaction-type diversity on community stability in these models, we revealed that sign-based and fitness-component-based interaction diversities contribute to community stability and its relationship with network complexity in different ways.

Interaction-type diversity resulting from the mixing of two interaction types enhances community stability only when it is both sign- and fitness-component-based (**I**). In the absence of fitness-component-based diversity (**II**), the community diversity monotonically decreases as the proportion of mutualism increases, supporting the previous findings of Suweis et al. (2014). In the absence of sign-based diversity (**III** and **IV**), community stability is maximized when there is no diversity in interaction type, in a similar way to the model without fitness-component-based diversity (**II**), suggesting that sign-based diversity is necessary for interaction-type mixing to stabilize a community. Thus, a combination of both fitness-component-based and sign-component-based interaction-type diversity is required to stabilize a community.

The interaction-type diversity that reverses the otherwise negative relationship between complexity and stability is sign based, with a positive complexity–stability relationship only emerging in the presence of sign-based interaction-type diversity (**I** and **II**). By contrast, in the absence of sign-based interaction-type diversity (**III** and **IV**), community stability is lowered or only slightly affected by an increased community complexity. In nature, mutualistic and antagonistic interactions often contribute to the same fitness component—for example, an animal seed disperser that helps plants to disperse their offspring receives nutrition or energy from the seed as a reward (mutualism) while also gaining nutrition or energy by eating other plant materials (antagonism). In such cases, the fitness component is identical for the mutualistic and antagonistic interactions, suggesting that this type of interaction-type diversity can give rise to a positive complexity–stability effect. By contrast, the coexistence of different kinds of mutualism (or antagonism) is unlikely to result in a positive complexity–stability relationship.

8.2.2 Diverse Ways of Allocating Interaction Efforts

There are various types of interaction effort allocations, as considered in models **I** and **II** above. In model **I**, interaction efforts are separately allocated to different interaction types (separated effort allocation) while the total interaction effort is kept constant (mixed effort allocation). These represent the two extremes of the ways in which interaction efforts can be allocated. In natural ecosystems, diverse species that allocate interaction efforts in a variety of ways are likely to coexist. Therefore, Kondoh and Mougi (2015) relaxed these extreme assumptions to examine whether the ITDH is applicable in a wider ecological context.

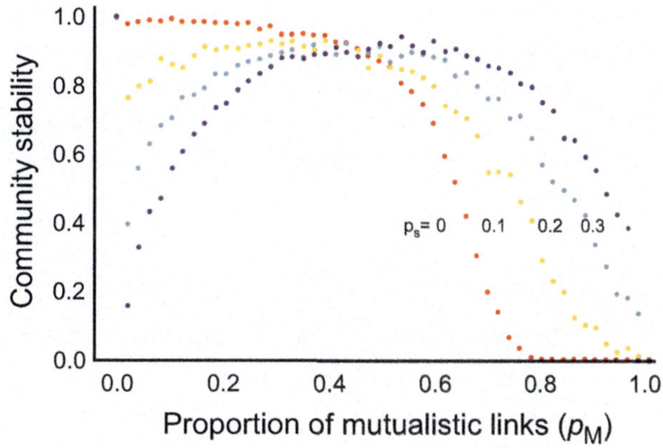

Fig. 8.5 Relationships between the proportion of mutualistic links (p_M) and stability with varying proportions of species for which antagonistic and mutualistic interaction efforts are rescaled separately

Consider a hybrid community with both antagonism and mutualism, within which species with separated and mixed effort allocations coexist at proportions of p_s and $1-p_s$, respectively. The type of effort allocation alters the relationship between community stability and interaction-type diversity (Fig. 8.5). The most prominent difference is observed when mutualism is rare (low p_M). When the effort allocation is of a mixed type ($p_s = 0$), community stability is highest at $p_M = 0$ and monotonically decreases with increasing p_M, in agreement with the earlier study of Suweis et al. (2014). However, when the effort allocation is of a separate type ($p_s = 1$), community stability is high at $p_M = 0$ and sharply declines at low p_M, following which it increases to reach a maximum at an intermediate p_M and then decreases to the lowest value at $p_M = 1$ (Fig. 8.3).

So what relationship emerges at an intermediate p_s? The decline in stability at a low p_M is observed under a wide range of p_s values (Fig. 8.5). However, a monotonic decline in stability with increasing p_M is only observed under the extreme setting where all of the effort allocation is of the mixed type ($p_s = 0$), with only a small proportion of separate effort allocation ($p_s > 0.1$) being required for the unimodal pattern to be revived.

The relationship between complexity and stability is also surprisingly robust to changes in the p_s value. Indeed, even if the effort allocation is of the mixed type ($p_s = 0$), the stability maximum emerges at the highest complexity only in the presence of interaction-type diversity, as shown by Suweis et al. (2014).

8.2.3 Interaction Effort Allocations Play a Key Role in Large Ecological Networks

The difference between classical models (May 1972; Allesina and Tang 2012) and the present models (**I** and **II**) is related to the assumption around how the interaction strength depends on the numbers of links per species. In the earlier models, the average effect size of interspecific interactions was assumed to be independent of both species richness and connectance. However, I would argue that those models made unrealistic assumptions—for example, that a species can get an infinitely large fitness gain in an infinitely large mutualistic community. Therefore, it may not be appropriate to apply their model predictions directly to large, complex ecological communities.

8.2.4 Roles of Overlooked Unilateral Interactions in Community Stability

In the above sections, I demonstrated that the diversity of interaction types could play a key role in stabilizing ecological communities, with a specific consideration of three major interaction types: antagonism, mutualism, and competition. These interactions commonly have reciprocal effects on the interacting partners. However, other potentially common interactions in natural communities are not reciprocal but unilateral (one species has a marked effect on only one other species). Therefore, it remains unclear how such unilateral interactions affect the stability of community dynamics.

Many interactions are strongly asymmetric (one interaction in a pairwise interaction is strong, while the other is weak). The two extremes of unilateral interactions are amensalism and commensalism. Amensalism can be defined as an interaction in which one organism inflicts harm on another organism without receiving any costs or benefits—for example, large terrestrial vertebrates crushing grass and small terrestrial invertebrates as they move. By contrast, commensalism can be defined as an interspecific relationship in which one species (the commensal) obtains benefits such as food, shelter, or locomotion from another species (the host) without causing adverse effects. Commensalism often occurs between a larger host and a smaller commensal—for example, remoras that attach to sharks and other fishes, and various biting lice, fleas, and mosquitoes that feed harmlessly on the skin of mammals. Antagonism (+,−), such as prey–predator and host–parasite relationships, may be amensalistic (0,−) or commensalistic (+,0), and competition (−,−) may also be amensalistic, while an interaction that is mutualistic (+,+) in one ecological setting could be commensalistic in another.

The inherent diversity of unilateral interactions suggests that elucidating the role of unilateral interactions in community dynamics will provide an understanding of whole communities rather than parts of a community. Therefore, to reveal the effects of unilateral interactions on community stability, I controlled the asymmetry of interaction strengths in communities with a single interaction type and the composition of unilateral interactions within a whole community with all interaction types (Mougi 2016a).

Consider a community with reciprocal antagonistic, competitive, or mutualistic interactions and unilateral amensalistic or commensalistic interactions. The proportion of unilateral interactions within the community is defined as p_u, and the proportion of reciprocal interactions is defined as $1 - p_u$. The population dynamics and the effect of unilateral interactions on the stability of the population dynamics can be evaluated by systematically changing p_u (Mougi 2016a).

I found that where only reciprocal interactions occur ($p_u = 0$), communities with antagonism are more stable than those with mutualism or competition, supporting the findings of Allesina and Tang (2012). However, increasing the asymmetry of the interaction strengths (which moves the communities toward unilateral interactions, i.e., amensalism and/or commensalism) decreases the stability in antagonistic communities while increasing the stability in competitive and mutualistic communities. These results suggest that asymmetrical, unilateral interactions have a more stabilizing effect than symmetrical, reciprocal interactions such as competition and mutualism but a more destabilizing effect than asymmetric, reciprocal interactions, such as predator–prey or host–parasite interactions.

Next, consider another extreme case where all of the species interactions are completely asymmetrical or unilateral ($p_u = 1$). Communities that are skewed toward one type of unilateral interaction type are unstable, whereas those with a moderate mix of amensalism and commensalism tend to be more stable. This unimodal pattern of stability is observed at intermediate levels of the parameters (Mougi 2016a), suggesting that a mixing of unilateral interactions has a tendency to enhance community stability when all of the parameters have moderate variations.

Finally, consider a community with all interaction types ($0 < p_u < 1$). Regardless of the proportion of each interaction type, communities with unilateral interactions tend to be more stable than those with reciprocal interactions. Furthermore, unilateral interactions can compensate for unstable reciprocal interactions. This stabilization is caused by the zero interaction strengths of unilateral interactions decreasing the connectance by half and supports the predictions of earlier theories (Allesina and Tang 2012).

8.3 Role of Adaptation in Hybrid Communities

So far, I have considered static interaction networks, following the approach of classical studies. However, the interaction network can change over time in terms of the links and strengths. Adaptation has been proposed as a key driver of such nonstatic networks, as it inherently alters species interactions and results in changes in the population dynamics. Kondoh (2003) demonstrated that a behavioral or evolutionary adaptive shift in partner by foragers greatly changes the food-web structure and dynamics and creates a positive complexity relationship, leading to the hypothesis that the flexibility of food webs that were not assumed in classical static interaction network systems is a key consideration for resolving May's paradox (May 1972). Similarly, later studies have also demonstrated that adaptive partner shifts in mutualistic interactions can shape the network structures of mutualistic communities, which may promote species coexistence (Zhang et al. 2011).

However, although these findings suggest that adaptation can play a critical role in the persistence of large, complex communities, it remains unclear how adaptation works in a hybrid community with multiple interaction types.

8.3.1 Adaptation in a Hybrid Module

Let us begin by considering a simple module. A widespread example of a community module in which exploiters and mutualists share interacting species is that of terrestrial plant and animal species. The animal species of interest may be a mutualist, such as a pollinator or seed disperser, or an exploiter, such as an herbivore, seed feeder, or leaf miner. There is evidence that animal exploiters and mutualists adaptively switch between resource plants (Egas and Sabelis 2001; Whittall and Hodges 2007). The module, which consists of a resource, its mutualist, and an exploiter, is the simplest building block of a community in a terrestrial ecosystem.

Consider two species that compete for resources asymmetrically and cannot coexist on their own because the superior species always excludes the inferior species. Mougi and Kondoh (2014c) explored how the presence of an adaptive exploiter and/or adaptive mutualist affects the competitive outcome and persistence of the whole system. For adaptation, the interaction effort was allocated among potential resources, as was assumed in earlier studies (Kondoh 2003). More specifically, the interaction effort among resource species was reallocated among interacting partners in a way that increased fitness based on a constraint that the efforts sum to unity. This was achieved by increasing (or decreasing) the interaction effort allocations to the resource species that had a higher (or lower) energy gain per unit effort than the average.

It was found that the equilibrium species composition in a system with adaptive switching depends on the relative strength of the antagonistic and mutualistic interactions (S) and can be classified into three phases. When S is small, the exploiter does not persist and the mutualist chooses the superior competitor (or more abundant species) as its partner, resulting in mutualist-driven competitive exclusion because the superiority of the superior competitor is enhanced. When S is larger, the inferior competitor is again excluded while the other species coexists. However, when S is larger still, all four species coexist. This pattern suggests that a larger value of S is required for the four species to coexist. This is because adaptive switching of the exploiter effectively regulates the instability caused by the adaptive mutualist when the exploitation is strong. However, this does not mean that mutualism is not also necessary for the coexistence of these species—if the mutualist has an indirect positive effect on the exploiter, it will greatly contribute to community persistence, and this indirect effect may also allow a nonpersistent subsystem that is comprised of a weak exploiter and two resources (the inferior resource goes extinct) to persist in the absence of a mutualist.

Both adaptive antagonism and adaptive mutualism can interactively support the coexistence of multiple species. While adaptive antagonism can potentially enhance the coexistence of competing resources, the interaction needs to be strong enough for this predator-mediated coexistence mechanism to work—thus, a community that includes two competing species and an adaptive exploiter with weak interactions will not be able to persist. However, this nonpersistent community can persist if an adaptive mutualist is added, which potentially inhibits resource coexistence in the absence of an adaptive exploiter. This pattern demonstrates the critical role that the coexistence of multiple interaction types plays in community maintenance.

The mediation of species coexistence by antagonism and mutualism can be explained as follows. The adaptive exploiter favors the more abundant resource species, placing this resource at a disadvantage. Thus, its presence tends to prevent competitive exclusion at the resource level. For the same reason, a "weak" exploiter cannot allow competing resource species to coexist. However, this predator-mediated coexistence mechanism can be enhanced by the presence of a mutualist, as adaptive mutualism effectively supports the adaptive exploiter via a positive indirect effect that is mediated by the resource species and strengthens the exploitation, which can regulate the superior competitor and enhances coexistence.

Adaptation has a key effect on community stability, tending to enhance species coexistence in resource–exploiter systems (Kondoh 2003) and inhibiting species coexistence in mutualistic systems. When the mutualistic interaction is relatively strong or the antagonistic interaction is relatively weak, the effects remain unchanged: adaptations in antagonism and mutualism enhance and inhibit coexistence, respectively, even in a mixed community. However, when the mutualistic interaction is relatively weak, the adaptation effects can be reversed. Faster adaptation of the mutualist enhances species coexistence by stabilizing the community dynamics, whereas faster adaptation of the exploiter may enhance or prevent coexistence depending on the speed of mutualist adaptation. Therefore, it is not possible to determine whether adaptation will promote or inhibit species coexistence without knowing the relative interaction strength and adaptation speed of the different interaction types.

These results suggest that adaptation and interaction types synergistically contribute to community persistence. However, it remains unclear whether these synergistic effects hold in larger, more complex systems.

8.3.2 Adaptation in Large Hybrid Networks

Here, I discuss how adaptive partner shifts in both antagonistic and mutualistic interactions affect the stability of the dynamics of complex communities. To examine the effects of adaptation and interaction-type diversity on community stability with a given community complexity, I considered a bipartite network as a conceivable realistic network by merging both antagonistic and mutualistic interactions and controlled the adaptation speed (G) and the proportion of mutualism (p_M) within a community (Mougi 2016b).

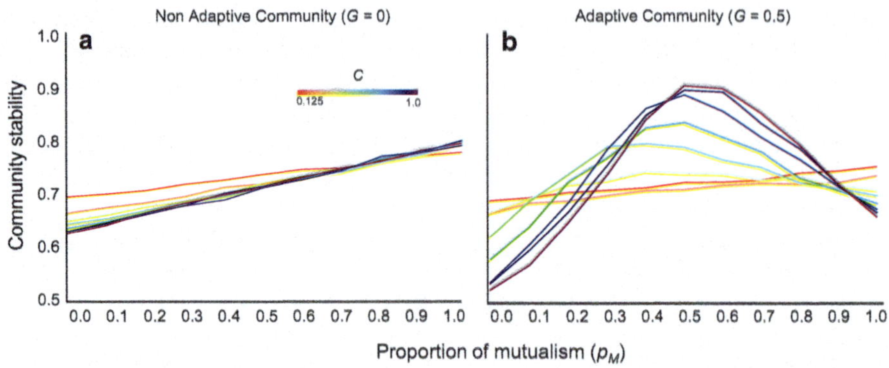

Fig. 8.6 Relationships between the proportion of mutualistic links (p_M) and stability in a community (**a**) without adaptation ($G = 0$) and (**b**) with adaptation ($G > 0$). Different colors indicate different levels of connectance (C). Modified from Mougi (2016b). © 2016. Nature Publishing Group

First, consider communities that are comprised of a single interaction type ($p_M = 0$ or 1). In both types of communities, adaptation tends to destabilize the population dynamics (Fig. 8.6b). However, the instability is higher when community complexity (e.g., C) is higher, indicating that there is a negative relationship between community complexity and stability in the same way as occurs with no adaptation (Fig. 8.6). However, contrary to earlier theories (Kondoh 2003), adaptation causes instability in nonhybrid communities. In the random or cascade food webs that were assumed by Kondoh (2003), adaptive exploiters are always likely to utilize more abundant species, resulting in the rescue of rare species and a stabilizing effect. However, in a bipartite network, rare resource species are unlikely to be rescued for three reasons: (1) the consumers' predation pressures on the resource species are inherently higher than in random and cascade networks owing to the absence of top predators; (2) interspecific competition between the resource species promotes the extinction of rare resource species that are utilized by consumers; and (3) partner shifts after the extinction of rare resource species increase the predation pressures on alternative resource species. Mutualists also have a destabilizing effect, as mentioned previously, as adaptive mutualists favor the more abundant resource species, providing a mechanism of majority advantage and thus potentially inhibiting the coexistence of resource species.

The mixing of both interaction types within a community dramatically changes the effects of adaptation on community dynamics in three ways. First, adaptation creates a stabilizing effect due to the hybridity of multiple interaction types. Although the hybridity of interaction types does not qualitatively change community stability in the absence of adaptation (Fig. 8.6a), a hybrid community has higher stability than a nonhybrid community in the presence of adaptation, reaching a peak at an intermediate mixing of different interaction types (Fig. 8.6b). Second, faster adaptation enhances the stability of a hybrid community compared with a nonhybrid community (Mougi 2016b). Third, adaptation can reverse the otherwise negative complexity–stability relationship of communities to produce a positive relationship (Mougi 2016b) as long as different interaction types are mixed at intermediate levels.

The stabilizing effect of adaptation can be explained by the following mechanism. Adaptive exploiters rapidly begin to utilize abundant resource species that have benefited from mutualists, thereby avoiding the competitive exclusion of less abundant resource species. Adaptive mutualists then respond to the resulting reduction in previously abundant resources by rapidly shifting to other resources, thereby promoting the rapid recovery of resource species. As shown above, although this stabilization mechanism works in a simple hybrid module, it will be more efficient in more complex communities. These results suggest that the diversity of interaction types plays a critical role in maintaining communities containing adaptive organisms.

8.4 Conclusion

In this chapter, I argue that the diversity of interaction types is a key factor that stabilizes larger, more complex communities, particularly when there is some allocation of interaction efforts. The potential mechanism of effort allocation is adaptive partner shifts.

References

Allesina S, Tang S (2012) Stability criteria for complex ecosystems. Nature 483:205–208

Allesina S, Tang S (2015) The stability-complexity relationship at age 40: a random matrix perspective. Pop Ecol 57:63–75

Bascompte J, Jordano P, Melian CJ, Olesen JM (2003) The nested assembly of plant–animal mutualistic networks. Proc Natl Acad Sci U S A 100:9383–9387

Egas M, Sabelis MW (2001) Adaptive learning of host preference in a herbivorous arthropod. Ecol Lett 4:190–195

Elton CS (1958) Ecology of invasions by animals and plants. Chapman & Hall, London

Emmerson M, Raffaelli D (2004) Predator-prey body size, interaction strength and the stability of real food web. J Animal Ecol 73:399–409

Kéfi S, Berlow EL, Wieters EA, Navarrete SA, Petchey OL, Wood S et al (2012) More than a meal. Integrating non-feeding interactions into food webs. Ecol Lett 15:291–300

Kondoh M (2003) Foraging adaptation and the relationship between food web complexity and stability. Science 299:1388–1391

Kondoh M (2008) Building trophic modules into a persistent food web. Proc Natl Acad Sci U S A 105:16631–16635

Kondoh M, Mougi A (2015) Interaction-type diversity hypothesis and interaction strength: the condition for the positive complexity–stability effect to arise. Popul Ecol 51:21–27

Kondoh M, Mougi A (2017) What kind of interaction-type diversity matters for community stability? In: Moore JC, de Ruiter PC, McCann KS, Wolters V (eds) Adaptive food webs: stability and transitions of real and model ecosystems. Cambridge University Press, Cambridge, pp 19–30

Lafferty KD, Allesina S, Arim M, Briggs CJ, De Leo G, Dobson AP, Dunne JA, Johnson PT, Kuris AM, Marcogliese DJ, Martinez ND, Memmott J, Marquet PA, McLaughlin JP, Mordecai EA, Pascual M, Poulin R, Thieltges DW (2008) Parasites in food webs: the ultimate missing links. Ecol Lett 11:533–546

Lawlor LR (1978) A comment on randomly constructed model ecosystems. Am Nat 112:445–447

MacArthur R (1955) Fluctuations of animal populations, and a measure of community stability. Ecology 36:533–536

May RM (1971) Stability in multi-species community models. Math Biosci 12:59–79

May RM (1972) Will a large complex system be stable? Nature 238:413–414

May RM (1973) Qualitative stability in model ecosystems. Ecology 638:1389–1391

McCann KS (2000) The diversity-stability debate. Nature 405:228–233

Melián CJ, Bascompte J, Jordano P, Křivan V (2009) Diversity in a complex ecological network with two interaction types. Oikos 118:122–130

Mitani N, Mougi A (2017) Population cycles emerging through multiple interaction types. R Soc Open Sci 4:170536

Mougi A (2016a) The roles of amensalistic and commensalistic interactions in large ecological network stability. Sci Rep 6:29929

Mougi A (2016b) Stability of an adaptive hybrid community. Sci Rep 6:28181

Mougi A, Kondoh M (2012) Diversity of interaction types and ecological community stability. Science 337:349–351

Mougi A, Kondoh M (2014a) Instability of a hybrid module of antagonistic and mutualistic interactions. Popul Ecol 56:257–263

Mougi A, Kondoh M (2014b) Stability of competition-antagonism-mutualism hybrid community and the role of community network structure. J Theor Biol 360:54–58

Mougi A, Kondoh M (2014c) Adaptation in a hybrid world with multiple interaction types: a new mechanism for species coexistence. Ecol Res 29:113–119

Neutel A-M, Heesterbeek JAP, de Ruiter PC (2002) Stability in real food webs: weak links in long loops. Science 296:1120–1123

Odum EP (1953) Fundamentals of ecology. Saunders, Philadelphia

Pocock MJO, Evans DM, Memmott J (2012) The robustness and restoration of a network of ecological networks. Science 335:973–977

Sauve AMC et al (2014) Structure–stability relationships in networks combining mutualistic and antagonistic interactions. Oikos 123:378–384

Stouffer DB, Bascompte J (2011) Compartmentalization increases food-web persistence. Proc Natl Acad Sci 108:3648–3652

Suweis S, Grilli J, Martian A (2014) Disentangling the effect of hybrid interactions and of the constant effort hypothesis on ecological community stability. Oikos 123:525–532

Thébault E, Fontaine C (2010) Stability of ecological communities and the architecture of mutualistic and trophic networks. Science 329:853–856

Toju H, Guimarães PR, Olesen JM, Thompson JN (2014) Assembly of complex plant–fungus networks. Nat Commun 5:5273

Vance PR (1985) The stable coexistence of two competitors for one resource. Am Nat 126:72–86

Whittall JB, Hodges SA (2007) Pollinator shifts drive increasingly long nectoar spurs in columbine flowers. Nature 447:706–709

Zhang F, Hui C, Terblanche JS (2011) An interaction switch predicts the nested architecture of mutualistic networks. Ecol Lett 14:797–803

Part IV
A Novel Approach in Community Ecology

Chapter 9
Forecasting Ecological Time Series Using Empirical Dynamic Modeling: A Tutorial for Simplex Projection and S-map

Masayuki Ushio and Kazutaka Kawatsu

Abstract Natural ecosystems are often complex, dynamic and state-dependent (i.e., nonlinear), and it is difficult to forecast their (near) future states if we rely only on linear statistical approaches. In the past few decades, tools of nonlinear time series analysis have been developed to analyze and forecast the state-dependent behavior of nonlinear systems. These methods do not assume any set of equations governing the system, and thus are suitable for analyzing systems which are complex and for which it is therefore often difficult to make reasonable assumptions about their underlying mechanisms. Instead of assuming equations, these methods recover the dynamics (and potentially, their underlying mechanism) directly from time series data, and are thus called Empirical Dynamic Modeling (EDM). In this chapter, we will introduce a basic concept in EDM (i.e., state space reconstruction) and explain practical and detailed algorithms of two core EDM tools (i.e., simplex projection and S-map) to forecast ecological time series. Then, we show applications of this algorithm to community ecology and its potential to answer ecological questions.

Keywords Delay embedding · Empirical dynamic modeling · Forecasting · Nonlinear dynamics · Simplex projection · S-map

9.1 Introduction

Natural ecosystems consist of complex components, including many biotic/abiotic ones as well as their interactions. One consequence of such complex interactions is complex dynamics that are often difficult to understand intuitively. For example, in natural ecosystems, the signs and magnitudes of correlations between two variables

M. Ushio (✉)
Hakubi Center, Kyoto University Kyoto, Japan/Center for Ecological Research, Kyoto University, Kyoto, Japan

K. Kawatsu
Graduate School of Life Sciences, Tohoku University, Sendai, Japan

A. Mougi (ed.), *Diversity of Functional Traits and Interactions*, Theoretical Biology,
https://doi.org/10.1007/978-981-15-7953-0_9

can change with time, and thus predictions using linear extrapolations can perform only poorly in nonlinear systems (e. g., Sugihara et al. 2012; Ye et al. 2015a). Unfortunately, but importantly, such state-dependent behaviors are often found in natural ecosystems (e.g., Sugihara and May 1990a; Hsieh et al. 2005).

Time series can be defined as any set of sequential observations of the system state, and observation/measurement processes can be regarded as a function that maps the state onto the real number line. If a deterministic rule governs the system, the dynamic behaviors can be delineated as a trajectory of a state over time in a multidimensional state space (i.e., a manifold) by plotting time series with sufficient quality (e.g., constant observation intervals, reasonably frequent observations, and small observation errors) (see animation in https://youtu.be/fevurdpiRYg). Time series taken from ecosystems (i.e., ecological time series) can be used to trace out trajectories of the system, which provide a large amount of information on ecosystem dynamics. For example, if one has performed sequential observations on a three-species ecological system, e.g., grasses (primary producer), rabbits (prey), and foxes (predator), then the dynamics of the three-species system can be reconstructed by plotting time series of grasses, rabbits, and foxes along the x, y, and z axis, respectively, in a three-dimensional state space. The motion of the three-dimensional vectors indicates the system behavior.

In a natural ecosystem, however, it is almost impossible to collect time series of all potentially important variables involved in a target system (e.g., grasses, rabbits, and foxes in the three-species ecological example). Fortunately, Takens (1981) offered a theoretical basis to solve this problem: a mathematical theorem, Takens' embedding theorem, demonstrated that a shadow version of the attractor (motion of vectors in a state space) can be reconstructed by a single observed time series (for example, the time series of rabbits). In other words, delineation of trajectories, originally constructed using multivariable, can be possible even if a time series is available only for a single variable (Takens 1981; Sauer et al. 1991a). To embed such a single time series (with an equal sampling interval), vectors in the putative phase space are formed from time-delayed values of the time series, $\{x_t, x_{t-\tau}, x_{t-2\tau}..x_{t-(E-1)\tau}\}$, where E is the embedding dimension (i.e., the dimension or number of time-delayed coordinates required for the attractor reconstruction), and τ is the time lag (see Sugihara and May 1990a). This procedure, the reconstruction of the original dynamics, is known as State Space Reconstruction (SSR) (Takens 1981; Deyle and Sugihara 2011).

Takens' Theorem states that the shadow version of the dynamics reconstructed by SSR preserves the essential mathematical properties of the original system if sufficient lags are taken (Takens 1981). Importantly, local neighborhoods (and their trajectories) in the reconstruction map one-to-one to those of the original system. Even more importantly, points near each other in the reconstructed state space share similar past and future (i.e., evolve similarly in time), and thus the behavior of local neighbor points can be used to forecast the near future of a target point.

Recently developed tools for nonlinear time series analysis, which are specifically designed to analyze state-dependent behavior of dynamical systems, called Empirical Dynamic Modeling (EDM), are rooted in SSR (Sugihara and May 1990a;

Sugihara 1994a; Dixon et al. 1999; Deyle et al. 2013; Ye et al. 2015a, 2015b; Ye and Sugihara 2016; Ushio et al. 2018). These methods do not assume any set of equations governing the system, and thus are suitable for analyzing complex systems, for which it is often difficult to make reasonable *a priori* assumptions about their underlying mechanisms. Instead of assuming a set of specific equations, EDM recovers the dynamics (and potentially, underlying mechanism) directly from time series data, and is thus particularly useful for forecasting ecological time series, which are otherwise often difficult to forecast. Here we give step-by-step instructions on how to make predictions for ecological time series using two fundamental tools in EDM, i.e., simplex projection (Sugihara and May 1990a) and S-map (Sugihara 1994a). Simplex projection and S-map utilize the fact that points near each other in the reconstructed state space share similar past and future and enable accurate predictions of near future states of nonlinear dynamical systems. We demonstrate the algorithms using step-by-step R codes without using any existing EDM packages such as rEDM (Ye et al. 2015b) or multispatial CCM (Clark et al. 2015), which will help readers to understand the detailed procedure of EDM. All R codes and model data in this chapter are available at: https://github.com/ong8181/simplex-smap-tutorial.

9.2 Simplex Projection

9.2.1 Model Time Series

Simplex projection, developed by Sugihara and May (1990a), is one of the most fundamental tools in EDM. Simplex projection implements an algorithm that enables nonlinear predictions by leveraging information about the behavior of nearest neighbors of a target point in a state space. In this section, we will present the algorithm of simplex projection using the time series of a simple logistic map used in Sugihara et al. (2012). A similar explanation/example of simplex projection with a different model time series can be found in https://github.com/opetchey/RREEBES.

The logistic model contains two interacting species, X and Y, without measurement errors:

$$X(t+1) = X(t)[4 - 3X(t)]$$
$$Y(t+1) = Y(t)[3.1 - 3.1Y(t) - 0.31X(t)]$$

The first 500 time steps were discarded so that the dynamics leave transient states, and then the time series are plotted in Fig. 9.1. The model time series can be

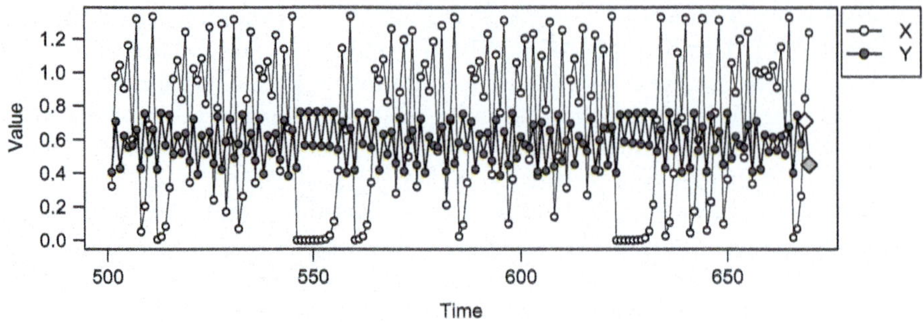

Fig. 9.1 Time series of the two-species logistic map. White and gray points represent the time series of X and Y, respectively. White and gray diamonds represent the target point and its one-time-step future predictee

downloaded from https://github.com/ong8181/simplex-smap-tutorial/blob/master/data/TwoSppUni_Example.csv, and we assume that the model time series is placed in "Rcode/data/" folder.

```
# Load model time series
d <- read.csv("Rcode/data/TwoSppUni_Example.csv")
tl <- nrow(d)
time <- d$t
# Usually data should be normalized (e.g., using scale() function) before EDM
.
# Here we do not perform normalization just for better visualization.
x <- d$X
y <- d$Y
```

The aim of simplex projection is to forecast the (near) future of the time series only by using the previous values of the same time series. Therefore, we will focus only on a single time series, Y, set a target point (i.e., the white diamond in Fig. 9.1), and try to forecast the one-time-step future value of Y (i.e., the gray diamond in Fig. 9.1) as a demonstration.

9.2.2 *Embedding*

Simplex projection starts with embedding the time series: we plot $Y(t)$ and $Y(t-1)$ in a two-dimensional state space (Fig. 9.2a).

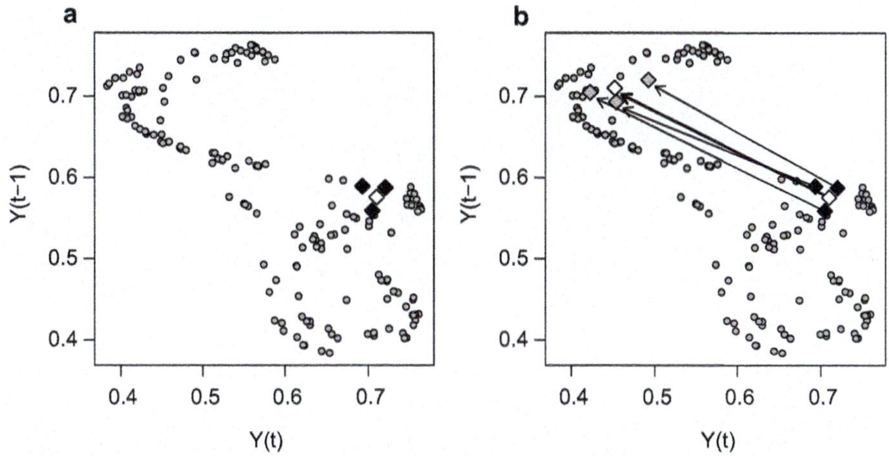

Fig. 9.2 Two-dimensional state space reconstructed using $Y(t)$ and $Y(t-1)$. The target point ($y(t^*)$) is indicated by the white diamond. (**a**) Three nearest neighbors are indicated by black diamonds. (**b**) One time-step forecasting based on the three nearest neighbors. White and gray diamonds indicated by arrows are the future points of the target and neighbors, respectively

```
# Assume that the best embedding dimension (E) = 2.
y.E <- 2
y.embed <- matrix(NA, nrow = length(y), ncol = y.E)
for(i in 1:y.E) y.embed[i:length(y), i] <- y[1:(length(y)-i+1)]

# Calculating Euclidean distances between the target and other points
set.target <- matrix(rep(y.embed[tl-1,], nrow(y.embed)), ncol = ncol(y.embed)
, byrow = T)
distances <- sqrt(rowSums((y.embed - set.target)^2))
neighbors <- order(distances[-(tl-1)])[1:(y.E+1)] # Exclude the target point
```

Simplex projection usually looks for $E + 1$ nearest neighbors in the state space (Sugihara and May 1990a). Following this procedure, we select three nearest neighbors based on the Euclidean distance between the target point and the other points. The target point and its three nearest neighbors are indicated by white and black diamonds, respectively, in Fig. 9.2a. What we did here was to find points that exhibit "similar behavior" with the target point. This can be more clearly illustrated if we go back to the original time series plot. As shown in Fig. 9.3a, finding nearest neighbors means finding points that show similar behavior (or trajectory) in the time series (black points in Fig. 9.3a).

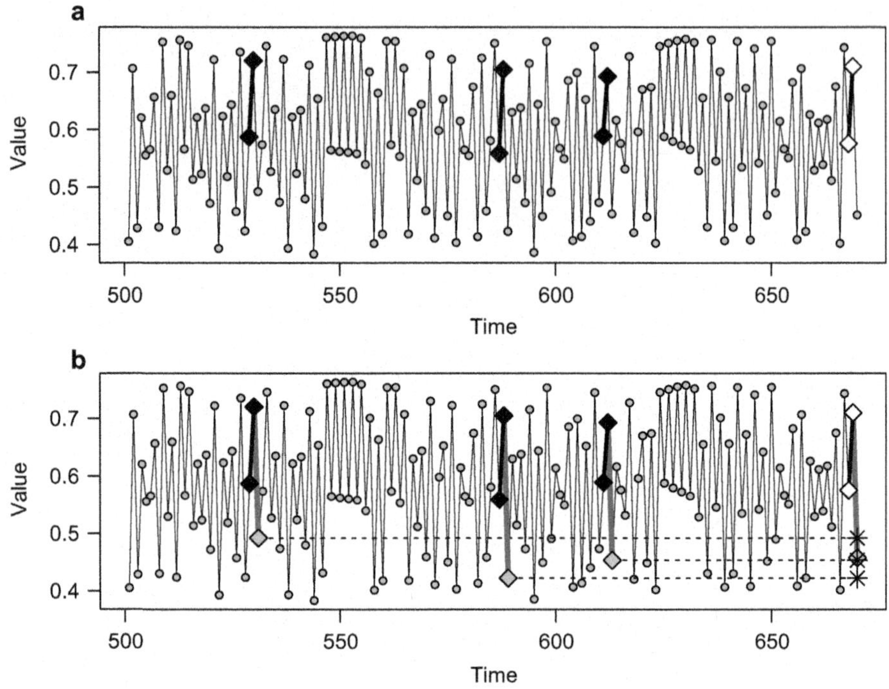

Fig. 9.3 Simplex projection explained using the original time series of the two-species logistic map. (**a**) The target point and three nearest neighbors are indicated by white and black diamonds, respectively. (**b**) The one-time-step future values of the target and neighbors are indicated by gray diamonds. The prediction is made based on the weighted average of the future values of the nearest neighbors (symbols, "*", indicate the future values of the nearest neighbors projected onto the time 670)

9.2.3 Calculating Weights

The next step is to calculate a weight for each nearest neighbor based on the Euclidean distance. The weight for the *i*th nearest neighbor, w_i, can be calculated as:

$$w_i = exp\left(\frac{-\left\|y(t^*) - y(t_i)\right\|}{\left\|y(t^*) - y(t_1)\right\|}\right)$$

where $\|x - y\|$ is the Euclidean distance between two points in an E dimensional space (i.e., $\|x - y\| = ([x_1 - y_1]^2 + [x_2 - y_2]^2 + \ldots + [x_E - y_E]^2)^{1/2}$), $y(t^*)$ is the target point at time t^*, and $y(t_i)$ is the *i*th nearest neighbor (i.e., $y(t_1)$ is the first nearest neighbor).

These calculations are implemented with R as follows:

```
# Compute weights
min_distance <- distances[neighbors[1]]
if(min_distance == 0){
  weights <- rep.int(0.000001, times = num_neighbors)
  weights[distances[neighbors] == 0] <- 1
}else{
  weights <- exp(-distances[neighbors]/min_distance)
  weights[weights < 0.000001] <- 0.000001
}
total_weight <- sum(weights)
```

9.2.4 Predictions

Once the identifications of nearest neighbors and calculations of weights have been done, the prediction of $Y(t^*)$ for p time steps into the future is:

$$Y(t^* + p) \mid M_y = \sum_{i=1}^{E} w_i Y(t_i + p) / \sum_{j=1}^{E} w_j$$

This can be simply implemented with R as follows:

```
# Perform one-step prediction
pred <- (weights %*% y[neighbors+1]) / total_weight
```

This prediction can be illustrated in the state space as shown in Fig. 9.2b. An alternative illustration of the prediction in the original time series plot may be more intuitively easy to understand (Fig. 9.3b): simplex projection predicts the future values based on the behavior of nearest neighbors that share similar past and future states with the target point.

These processes are iterated for all points in the state space, and predictions can be made for all points. Alternatively, one may divide the time series into halves, and use the first half to reconstruct the state space and the second half to test the performance of the prediction.

Next, R codes iterate the above process for all points in the state space. By performing predictions for all possible points, we can evaluate the forecasting skill of simplex projection by calculating Pearson's correlation between predicted and observed values (Fig. 9.4).

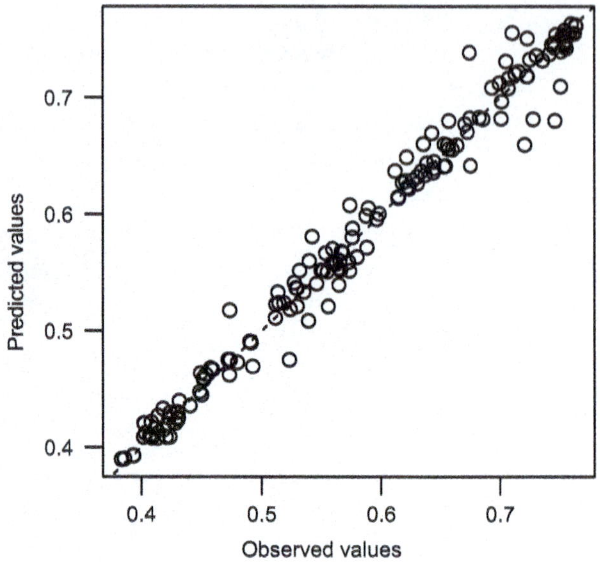

Fig. 9.4 Evaluation of forecasting skill using simplex projection. Dashed line indicates one-to-one line

```
# ts.d = time series data
# E = number of dimensions for the attractor reconstruction
# Alternatively, one may use a half of time series to reconstruct attractor.
ts.d <- y
E <- y.E
n <- length(ts.d)

# Make an output dataframe
output <- data.frame(index = 1:n, ts = ts.d, pred = rep(NA, n))

# Univariate SSR
ts.embed <- matrix(NA, nrow = n, ncol = E)
for(i in 1:E) ts.embed[i:n, i] <- ts.d[1:(n-i+1)]
```

```
# Calculate distances and make predictions
for(i in 1:(n-1)){
  if(!any(is.na(ts.embed[i,]))){ # skip if the vector contains NA
    set.target <- matrix(rep(ts.embed[i,], n), ncol = E, byrow = T)
    distances <- sqrt(rowSums((ts.embed - set.target)^2))
    neighbors <- order(distances)[-1][1:(E+1)] # Exclude the target point

    # Compute weights
    min.distance <- distances[neighbors[1]]
    if(min.distance == 0){
      weights <- rep.int(0.000001, times = num.neighbors)
      weights[distances[neighbors] == 0] <- 1
    }else{
      weights <- exp(-distances[neighbors]/min.distance)
      weights[weights < 0.000001] <- 0.000001
    }
    total.weight <- sum(weights)

    # One-step prediction
    output$pred[i+1] <- (weights %*% ts.d[neighbors+1]) / total.weight
  }
}

# Quantify the forecasting skill by Pearson's correlation coefficient
output <- output[complete.cases(output),]
cor(output$ts, output$pred, method = "pearson")
```

9.2.5 Estimating Best Embedding Dimension

Figure 9.4 illustrates how nearest neighbors in the state space share the near future of a target point. More importantly, a high forecasting skill suggests that the reconstructed attractor is smooth and well resolved. Thus, we may use the forecasting skill as a practical index of how well the dynamics are reconstructed (or how "smooth" the reconstructed dynamics are). By calculating a forecasting skill for different E using simplex projection, we can estimate E that can reasonably well reconstruct the original dynamics. In the R code below, simplex projection is performed using E from 1 to 10 (Fig. 9.5).

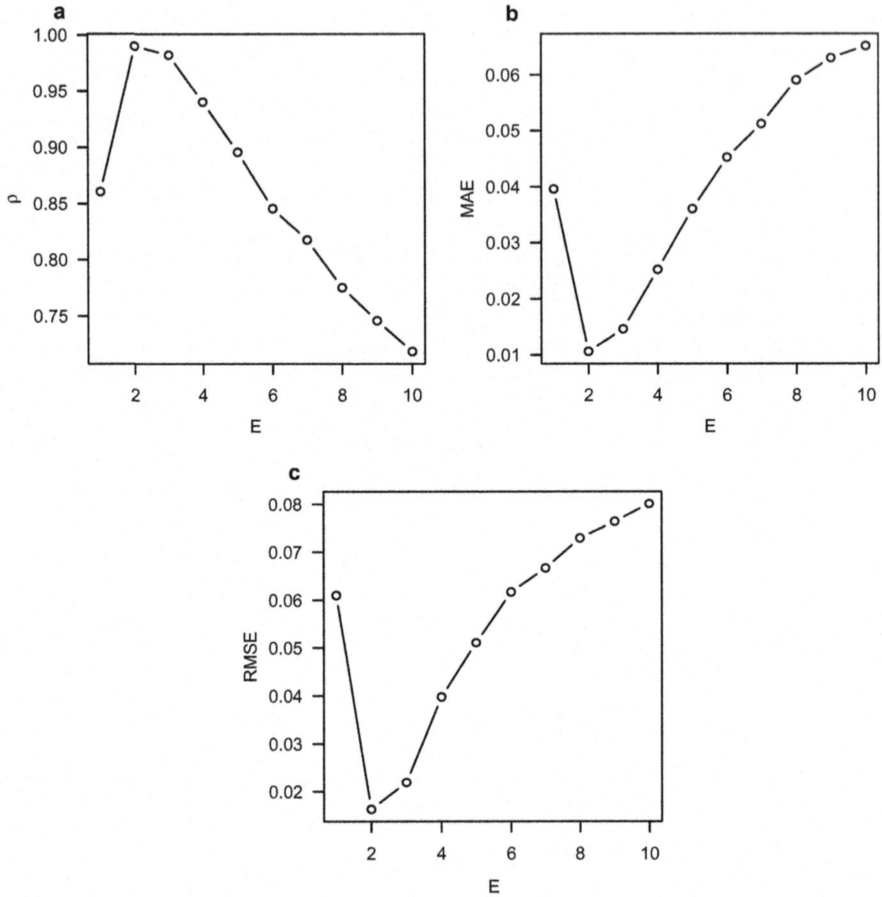

Fig. 9.5 Forecasting skill using simplex projection and embedding dimensions. Forecasting skills are evaluated by (**a**) correlation coefficients, (**b**) mean absolute error (MAE), and (**c**) root mean square error (RMSE)

```
# Evaluate forecasting skill for different E
rho.summary <- data.frame(E = 1:10, rho = NA, mae = NA, rmse = NA)

for(E.i in 1:10){
  # Simplex projection is implemented with the function "simplex.projection"
  # Definition can be found Rmarkdown file in GitHub
  simplex.res <- simplex.projection(y, E.i)
  simplex.res <- simplex.res[complete.cases(simplex.res),]
  rho.summary$rho[E.i] <- cor(simplex.res$ts, simplex.res$pred, method = "pea
rson")
  rho.summary$mae[E.i] <- mean(abs(simplex.res$pred - simplex.res$ts))
  rho.summary$rmse[E.i] <- sqrt(mean((simplex.res$pred - simplex.res$ts)^2))
}
```

The highest ρ (Pearson's correlation coefficient), the lowest MAE (mean absolute error), and the lowest RMSE (root mean square error) are all found at $E = 2$, suggesting that the attractor of $Y(t)$ is well resolved by using $Y(t)$ and $Y(t - 1)$ (i.e., the best embedding dimension is 2). Together, we demonstrated that simplex projection is a powerful tool to forecast ecological time series, and at the same time, it provides a practical way to estimate the best embedding dimension.

9.3 S-maps

The S-map (Sequential locally weighted global linear map) is the other fundamental tool in EDM developed by Sugihara (1994a). Similar to simplex projection, S-map also implements an algorithm that enables nonlinear forecasting, but it leverages information of all points in a state space (this is different from simplex projection, which uses only some nearest neighbors). We will demonstrate the algorithm of S-map using the identical time series used in the demonstration of simplex projection.

S-map also starts with embedding time series. The best embedding dimension, E determined by simplex projection, is usually used to reconstruct the state space. As seen in the above example, the best embedding dimension of $Y(t)$ was estimated to be 2, and thus we will use $E = 2$ in S-map.

9.3.1 Fitting a Locally Weighted Linear Model

First, a target point is specified in the same way as in simplex projection. For a target time point t^*, a linear model C is used to predict the future value $Y(t^* + p)$ from the points, $y(t^*)$, in the reconstructed state space ($y(t^*)$ has E elements). That is,

$$Y(t^* + p) = C_0 + \sum_{i=1}^{E} C_i y_j(t^*)$$

The linear model is fit to all other points in the state space. However, points that are close to the target point, $y(t^*)$, are given greater weights. Specifically, the model C is a solution obtained by singular value decomposition (SVD) to the equation:

$$B = AC$$

where B is an n-dimensional vector (n is from 1 to i) of the weighted future values $Y(t_i + p)$ for each historical time point, t_i, given by

$$B_i = w(i)Y(t_i + p)$$

and A is the $n \times E$ dimensional matrix (E is from 1 to j) given by

$$A_{ij} = w(i)y_j(t_i)$$

The weighing function w for t_i is defined by

$$w(i) = \exp\left(\frac{-\theta d}{d_m}\right)$$

which is the (Euclidean) distance between the target and ith points ($d = \|y(t_i) - y(t^*)\|$), and is tuned by the nonlinear parameter $\theta \geq 0$ and normalized by the average distance between the target, $y(t^*)$, and the other historical points, $y(t_i)$,

$$d_m = \frac{1}{n}\sum_{i=1}^{n} \left\| y(t_i) - y(t^*) \right\|$$

The situation in which the closer points are given greater weights can be illustrated in Fig. 9.6. In Fig. 9.6, points closer to the target (indicated by the white diamond) are given greater weights (indicated by the density of gray color).

Alternatively, we can go back to the original time series plot (an approach similar to that in Fig. 9.3). In Fig. 9.7, weights are indicated by the density of gray color and width of segments, meaning that points that show similar behavior with the target (white diamonds) are indicated by denser color.

In the S-map algorithm, points closer to the target point have more influence on the local linear model, as explained above, when θ is not zero. In other words, S-map with greater θ looks for a linear coefficient that makes residuals of closer points smaller (gray points in Fig. 9.8; points whose distance is smaller show smaller residuals). On the other hand, S-map with $\theta = 0$ (unweighted regression; equivalent to vector autoregressive model) looks for a linear coefficient that makes mean residuals smaller (white points in Fig. 9.8).

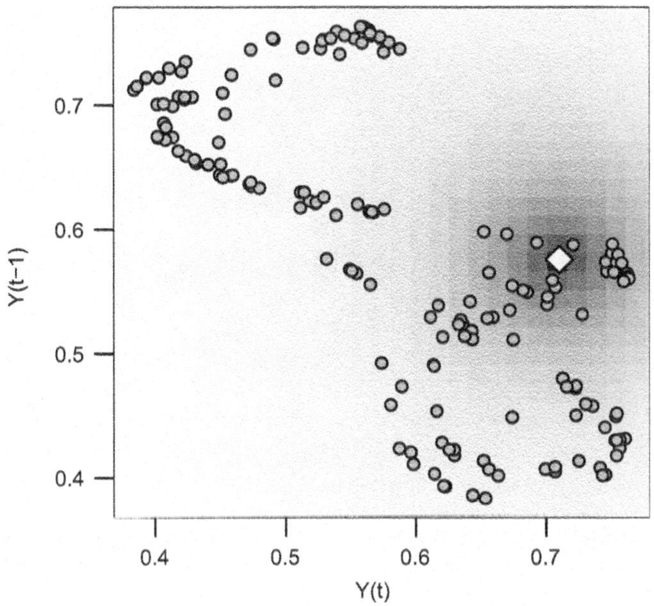

Fig. 9.6 Weights (θ) calculated for Euclidean distances between the target point ($y(t^*)$, white diamond) and the other points, indicated by the density of gray color

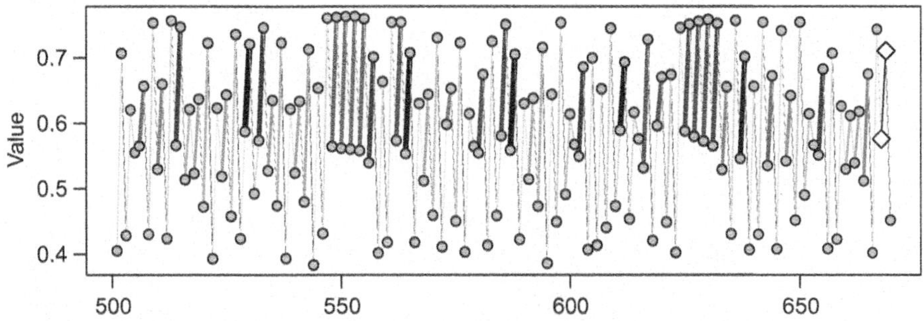

Fig. 9.7 Distance-dependent weights illustrated in original time series of the two-species logistic map. The densities of gray color and width of segments indicate the weight calculated based on the distances to the target points (indicated by white diamonds) in the state space

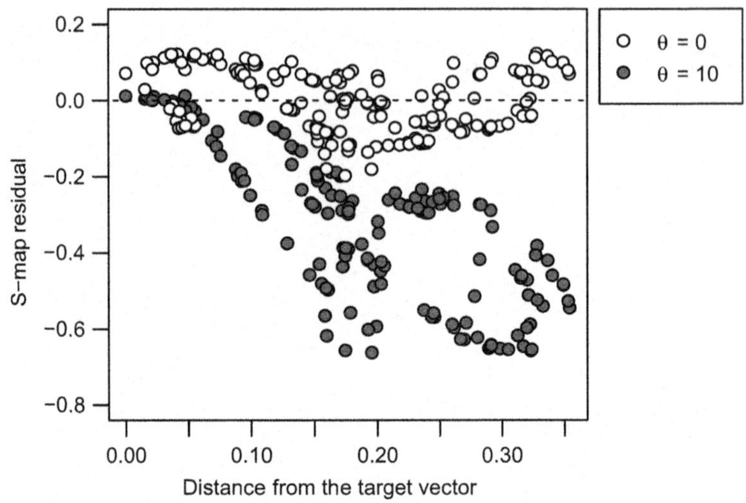

Fig. 9.8 Residuals of unweighted ($\theta = 0$) and weighted ($\theta = 10$) S-map. Note that the unweighted S-map is equivalent to the vector autoregressive model. White and gray points show residuals of the predictions using $\theta = 0$ and $\theta = 10$, respectively

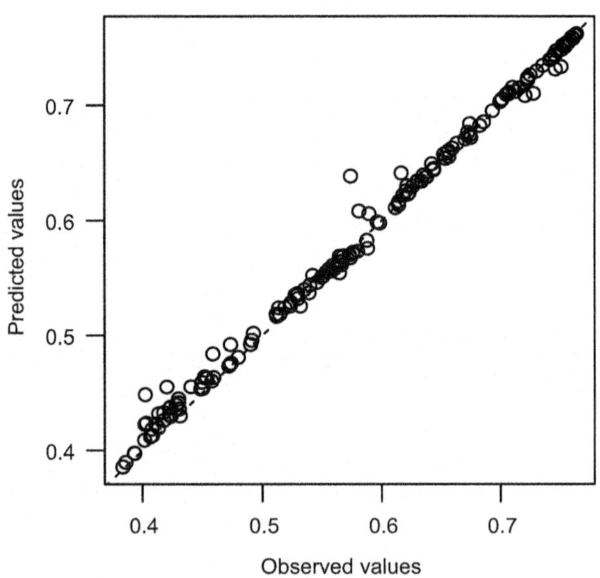

Fig. 9.9 Forecasting skill using S-map. Dashed line indicates one-to-one line. The forecasting skill using S-map is in general better than that using simplex projection (please compare Figs. 9.4 and 9.9)

9.3.2 Implementation of S-map with R

The algorithms explained above can be implemented with R as follows:

```
# ts.d = time series data
# E = number of dimensions for the attractor reconstruction
# Alternatively, one may use a half of time series to reconstruct attractor.
ts.d <- y
E <- y.E
n <- length(ts.d)



  theta <- 10

  # Make an output dataframe
  output <- data.frame(index = 1:n, ts = ts.d, pred = rep(NA, n))

  # Univariate SSR
  ts.embed <- matrix(NA, nrow = n, ncol = E)
  for(i in 1:E) ts.embed[i:n, i] <- ts.d[1:(n-i+1)]

  # Calculate distances and make predictions
  for(i in 1:(n-1)){
    if(!any(is.na(ts.embed[i,]))){ # skip if the vector contains NA
      set.target <- matrix(rep(ts.embed[i,], n), ncol = E, byrow = T)
      distances <- sqrt(rowSums((ts.embed - set.target)^2))

      # Compute weights
      d.m <- mean(distances, na.rm = T)
      ws <- exp(-theta * distances / d.m)

      # Make data frame w/o NA and target point
      ts.ws0 <- cbind(ts.embed, ws, matrix(rep(NA, n), ncol = 1))
      ts.ws0[1:(n-1),4] <- ts.embed[2:n,1]
      ts.ws <- ts.ws0[-i,]
      ts.ws <- ts.ws[complete.cases(ts.ws),]

      # Singular-value decomposition (excluding the target point)
      A <- cbind(ts.ws[,1:2], 1) * ts.ws[,3]
      A.svd <- svd(A)

      # Remove singular values that are too small
      s <- A.svd$d
      s.inv <- matrix(0, nrow = E+1, ncol = E+1)
      for(j in seq_along(s))
```

```
    {
      if(s[j] >= max(s) * 1e-5)
        s.inv[j,j] <- 1/s[j]
    }

      # Perform back-substitute to solve
      map <- A.svd$v %*% s.inv %*% t(A.svd$u) %*% (ts.ws[,3] * ts.ws[,4])

      # Make prediction
      output$pred[i+1] <- sum(map * c(ts.embed[i,], 1))
  }
}

# Quantify the forecasting skill
output <- output[complete.cases(output),]
cor(output$ts, output$pred, method = "pearson")
```

As shown in the above implementation, S-map performance depends on the weighting parameter, θ (preliminarily defined as 10 in the above implementation). As θ becomes larger, the local linear model utilizes more local information. If the forecasting skill improves with increasing θ, or reaches its maximum when $\theta > 0$, it is an indication that the dynamics being analyzed are state-dependent (i.e., a nonlinear process). On the other hand, if the forecasting skill reaches its maximum when $\theta = 0$, the S-map model is reduced to the Vector Autoregressive Model (VAR model). If forecasting skill is its maximum when $\theta = 0$, then the dynamics being analyzed are state-independent (i.e., a linear process). Therefore, θ can be used as a practical index to quantify the nonlinearity (Sugihara 1994a; Hsieh et al. 2005). Below is an example of quantification of the nonlinearity using S-map (Fig. 9.10).

```
# Evaluate forecasting skill for different theta
smap.summary <- data.frame(theta = 0:10, rho = NA, mae = NA, rmse = NA)

for(theta.i in 0:10){
  # S-map is implemented with the function "s.map"
  # Definition can be found Rmarkdown file in GitHub
  smap.res <- s.map(y, 2, theta.i)
  smap.res <- smap.res[complete.cases(smap.res),]
  smap.summary$rho[theta.i+1] <- cor(smap.res$ts, smap.res$pred, method = "pe
arson")
  smap.summary$mae[theta.i+1] <- mean(abs(smap.res$pred - smap.res$ts))
  smap.summary$rmse[theta.i+1] <- sqrt(mean((smap.res$pred - smap.res$ts)^2))
}
```

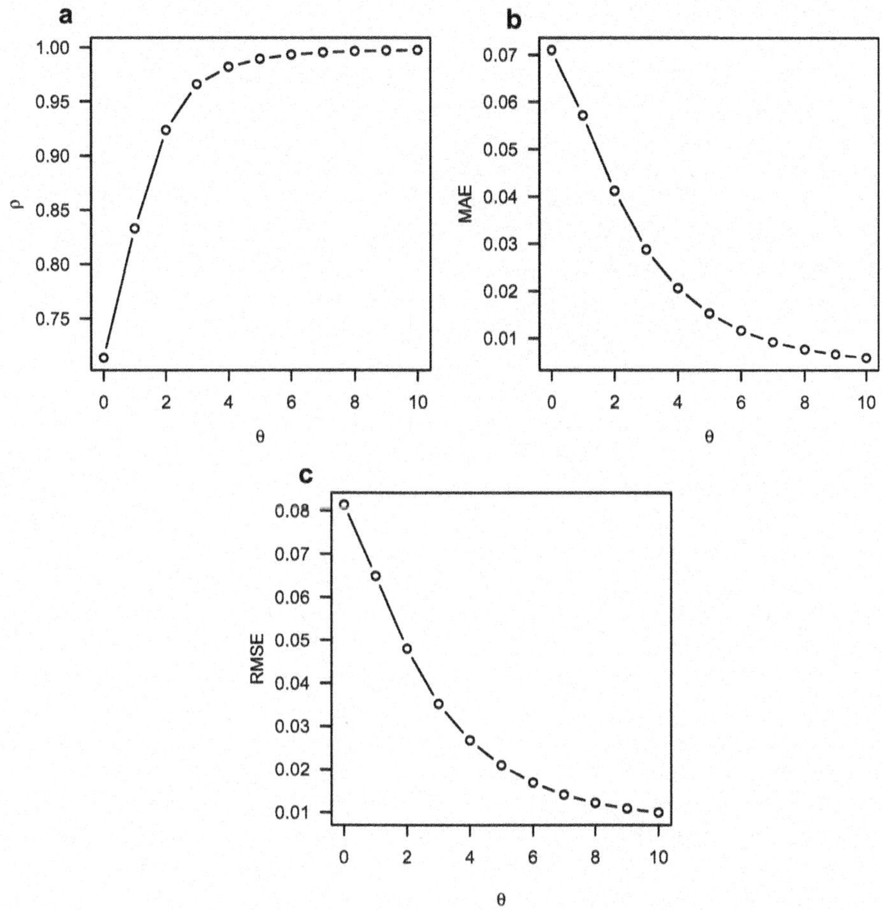

Fig. 9.10 Forecasting skill using S-map and nonlinear parameter θ. Forecasting skills are evaluated using (**a**) correlation coefficient, (**b**) mean absolute error (MAE), and (**c**) root mean square error (RMSE)

In the above example, the forecasting skill is best when $\theta = 10$, suggesting that the dynamics are nonlinear and state-dependent. Taking these considerations all together, S-map is also a powerful tool for making accurate predictions about nonlinear dynamic processes, and it also provides a practical way to quantify the nonlinearity of the dynamics.

9.4 Further Applications of Simplex Projection and S-map

Simplex projection and S-map are the two most fundamental tools in EDM (Sugihara and May 1990a; Sugihara 1994a). There are many further applications and extensions of these techniques. Here, we will just briefly introduce the applications and extensions because some review papers have already explained them more extensively (e.g., reviewed in Chang et al. 2017, Tsonis et al. 2018).

In terms of forecasting, the multivariate S-map/simplex projection is a more powerful approach (Deyle and Sugihara 2011; Deyle et al. 2013, 2016b). The multivariate S-map/simplex projection is a simple extension of S-map/simplex projection that incorporates multivariate time series in SSR (e.g., use of both *X*(*t*) and *Y*(*t*) in the above example). Usually, if the included time series have causal influences on the target time series (e.g., *Y*(*t*)), they will improve the forecasting skill. The multivariate S-map/simplex projection can be implemented by doing SSR using multivariate time series (e.g., R code example: ts.m.embed <- cbind(y,x); ts.m.embed can be used instead of ts.embed in the implementation).

Multi-view forecasting is an even more powerful method for forecasting (Ye and Sugihara 2016). The central idea of the multi-view forecasting is that consideration of multidimensional time series enables system dynamics to be reconstructed from multiple viewpoints (i.e., multiple embeddings). Ye and Sugihara (2016) selected several "best performing" embeddings and leveraged predictions from the multiple embeddings to make a better prediction. Indeed, in general, the multi-view forecasting enabled better predictions than univariate or multivariate SSR.

Another powerful application of these methods is convergent cross-mapping (CCM), which enables determinations of causal variables in nonlinear dynamical systems (Sugihara et al. 2012). A critical idea behind CCM is that, if two variables *X* and *Y* are part of the same dynamic system (i.e., they are causally related), then the reconstructed state spaces of *X* and *Y* will topologically represent the same attractor. The information of a causal variable, e.g., *X*, is included in time series of *Y*, and thus it is possible to predict the current state of a causal variable, *X*, from an effect variable, *Y*. The information of a causal variable in an effect variable can be looked for by testing whether there is a correspondence between their reconstructed state spaces (i.e., cross-mapping). Forecasting skill from one variable to another can be checked using simplex projection, and a significant forecasting skill is a practical criterion for the correspondence between two variables. In addition, the forecasting skill should increase with increasing time series length in SSR (or the number of points in a state space) if the two variables are causally related, which will be another practical criterion for the correspondence between two variables. CCM has recently been applied to many time series, and has shown its potential to detect causal relationships when manipulative experiments are not feasible (e.g., van Nes et al. 2015; Tsonis et al. 2015; Deyle et al. 2016a; Hannisdal et al. 2017; Nakayama et al. 2018; Ushio et al. 2018). For more details, please refer to some previous papers (Sugihara et al. 2012; Chang et al. 2017; Tsonis et al. 2018).

In ecological studies, one may be interested in mechanisms underlying community dynamics. EDM also provides approaches that can be used to infer mechanisms, e.g., interspecific interactions that drive the dynamics of natural ecosystems. For example, multivariate S-map allows quantification of fluctuating interspecific interactions (Deyle et al. 2016b; Kawatsu and Kishi 2017). In addition, a recent study showed that even fluctuating stability of a natural ecological community can be quantified and tracked using local Lyapunov exponents derived from fluctuating interspecific interactions (Ushio et al. 2018).

In addition, EDM is gaining attention in scientific fields even beyond ecology. For example, EDM has been applied to time series data in climate science (van Nes et al. 2015; Tsonis et al. 2015), epidemiology (Deyle et al. 2016a), neuroscience (Tajima et al. 2017), and paleoecology (Hannisdal and Liow Lee 2018). As multivariate time series in microbiome studies are accumulating, EDM also has the potential of being applied to such microbiome studies (Ushio 2020). Because inferring causal relationships, quantifying interaction strength, and forecasting when only time series are given is of general interest in many scientific fields, EDM will be gaining further attention and applied to more studies in many disciplines.

9.5 Concluding Remarks

In this review, we have provided an implementation of EDM without using any R packages. We have done this because we believe that implementing an EDM algorithm without packages is one of the best ways to learn how EDM works. In practice, however, we recommend using an R package "rEDM" (Ye et al. 2015a; Ye et al. 2018) as it implements EDM using RCpp packages, which is much faster than the R codes we provided. The latest version of "rEDM" (rEDM-1.5.0) can be downloaded either from CRAN (https://cran.r-project.org/web/packages/rEDM/). Example R codes using rEDM can be found in the supplement of Chang et al. (2017).

Obviously, one who wishes to understand EDM deeply needs to learn its theoretical background (e.g., embedding theorem; Takens 1981, Sauer et al. 1991a, Deyle and Sugihara 2011). However, we believe this review will help general ecologists to use powerful applications of EDM, which will facilitate forecasting and understanding the dynamics of natural ecosystems.

Acknowledgements This research is supported by PRESTO (JPMJPR16O2) from Japan Science and Technology Agency (JST). We would like to thank Hao Ye for his comments on the manuscript.

References

Chang C-W, Ushio M, Hsieh C-h (2017) Empirical dynamic modeling for beginners. Ecol Res 32:785–796

Clark AT, Ye H, Isbell F, Deyle ER, Cowles JM, Tilman D, Sugihara G (2015) Spatial 'convergent cross mapping' to detect causal relationships from short time-series. Ecology 96:1174–1181

Deyle ER, Sugihara G (2011) Generalized theorems for nonlinear state space reconstruction. PLoS One 6:e18295

Deyle ER, Fogarty M, Hsieh C-h, Kaufman L, MacCall AD, Munch SB, Perretti CT, Ye H, Sugihara G (2013) Predicting climate effects on Pacific sardine. Proc Natl Acad Sci U S A 110:6430–6435

Deyle ER, Maher MC, Hernandez RD, Basu S, Sugihara G (2016a) Global environmental drivers of influenza. Proc Natl Acad Sci 113:13081–13086

Deyle ER, May RM, Munch SB, Sugihara G (2016b) Tracking and forecasting ecosystem interactions in real time. Proc R Soc B Biol Sci 283:20152258

Dixon PA, Milicich MJ, Sugihara G (1999) Episodic fluctuations in larval supply. Science 283:1528–1530

Hannisdal B, Liow Lee H (2018) Causality from palaeontological time series. Palaeontology 61:495–509

Hannisdal B, Haaga KA, Reitan T, Diego D, Liow LH (2017) Common species link global ecosystems to climate change: dynamical evidence in the planktonic fossil record. Proc R Soc B Biol Sci 284:02170722

Hsieh C-H, Glaser SM, Lucas AJ, Sugihara G (2005) Distinguishing random environmental fluctuations from ecological catastrophes for the North Pacific Ocean. Nature 435:336–340

Kawatsu K, Kishi S (2017) Identifying critical interactions in complex competition dynamics between bean beetles. Oikos 127:553–560

Nakayama S-I, Takasuka A, Ichinokawa M, Okamura H (2018) Climate change and interspecific interactions drive species alternations between anchovy and sardine in the western North Pacific: detection of causality by convergent cross mapping. Fish Oceanogr 27:312–322

Sauer T, Yorke JA, Casdagli M (1991a) Embedology. J Stat Phys 65:579–616

Sauer T, Yorke JA, Casdagli M (1991b) Embedology. J Stat Phys 65:579–616

Sugihara G (1994a) Nonlinear forecasting for the classification of natural time series. Philos Trans R Soc A Math Phys Eng Sci 348:477–495

Sugihara G (1994b) Nonlinear forecasting for the classification of natural time series. Philos Trans R Soc A Math Phys Eng Sci 348:477–495

Sugihara G, May RM (1990a) Nonlinear forecasting as a way of distinguishing chaos from measurement error in time series. Nature 344:734–741

van Nes EH, Scheffer M, Brovkin V, Lenton TM, Ye H, Deyle E, Sugihara G (2015) Causal feedbacks in climate change. Nat Clim Chang 5:445–448

Sugihara G, May RM (1990b) Nonlinear forecasting as a way of distinguishing chaos from measurement error in time series. Nature 344:734–741

Sugihara G, May R, Ye H, Hsieh C-h, Deyle E, Fogarty M, Munch S (2012) Detecting causality in complex ecosystems. Science 338:496–500

Tajima S, Mita T, Bakkum DJ, Takahashi H, Toyoizumi T (2017) Locally embedded presages of global network bursts. Proc Natl Acad Sci 114:9517–9522

Takens F (1981) Detecting strange attractors in turbulence. In: Rand D, Young L-S (eds) Dynamical systems and turbulence. Springer, Cham, pp 366–381

Tsonis AA, Deyle ER, May RM, Sugihara G, Swanson K, Verbeten JD, Wang G (2015) Dynamical evidence for causality between galactic cosmic rays and interannual variation in global temperature. Proc Natl Acad Sci U S A 112:3253–2356

Tsonis AA, Deyle ER, Ye H, Sugihara G (2018) Convergent cross mapping: theory and an example. In: Tsonis AA (ed) Advances in nonlinear geosciences. Springer International Publishing, Cham, pp 587–600

Ushio M, Hsieh C-h, Masuda R, Deyle ER, Ye H, Chang C-W, Sugihara G, Kondoh M (2018) Fluctuating interaction network and time-varying stability of a natural fish community. Nature 554:360–363

Ushio M (2020) Idea paper: predicting culturability of microbes from population dynamics under field conditions. Ecol Res 35:586–590. https://doi.org/10.1111/1440-1703.12104

Ye H, Beamish RJ, Glaser SM, Grant SCH, Hsieh C-H, Richards LJ, Schnute JT, Sugihara G (2015a) Equation-free mechanistic ecosystem forecasting using empirical dynamic modeling. Proc Natl Acad Sci U S A 112:E1569–E1576

Ye H, Deyle ER, Gilarranz LJ, Sugihara G (2015b) Distinguishing time-delayed causal interactions using convergent cross mapping. Sci Rep 5:14750

Ye H, Sugihara G (2016) Information leverage in interconnected ecosystems: overcoming the curse of dimensionality. Science 353:922–925

Ye H, Clark A, Deyle E, Munch S, Cai J, Cowles J, Daon Y, Edwards A, Keyes O, Stagge J, Ushio M, White E, Sugihara G (2018) rEDM: applications of empirical dynamic Modeling from time series. https://doi.org/10.5281/zenodo.1935847

The manufacturer's authorised representative in the EU is Springer Nature Customer Service Centre GmbH, Europaplatz 3, 69115 Heidelberg, Germany. If you have any concerns regarding our products, please contact ProductSafety@springernature.com

Printed and bound by CPI Group (UK) Ltd, Croydon, CR0 4YY
24/07/2026
02175908-0001